Energy and Evolutionary Conflict

Neil W. Blackstone

Energy and Evolutionary Conflict

The Metabolic Roots of Cooperation

Springer

Neil W. Blackstone
Dept of Biological Sciences
Northern Illinois University
Dekalb, IL, USA

ISBN 978-3-031-06058-8 ISBN 978-3-031-06059-5 (eBook)
https://doi.org/10.1007/978-3-031-06059-5

This Springer imprint is published by the registered company Springer Nature Switzerland AG
The registered company address is: Gewerbestrasse 11, 6330 Cham, Switzerland

Preface

Britton Chance once said to me, perhaps revealingly, "hard-working mitochondria are good for you; it's the lazy ones that are dangerous." This idea, tracing its way back to pre-chemiosmotic work done in the 1950s, encapsulates the central theme of this book. As with many things, however, my development of this theme took longer and followed a more circuitous path. In fact, I prepared a draft of a book a number of years ago that I was revising after favorable reviews, but I was troubled because it lacked a unifying theme. It took several years of learning about photosynthetic symbioses through work on corals before I began to better understand why the "lazy ones" were dangerous and what this might mean for the evolution of cooperation. Throughout this lengthy genesis, numerous friends and colleagues have contributed to these ideas in ways both large and small. Many thanks to all.

Dekalb, IL, USA　　　　　　　　　　　　　　　　Neil W. Blackstone

Contents

Chapter 1
Introduction

It is perhaps not a startling observation that human society depends above all on cooperation. Society by definition involves the cooperative actions of many individuals. Societies that perform cooperative tasks well (e.g., contact tracing of individuals with COVID-19) are hailed as successful, while those that do not are scorned as failures [1]. Measures of trust, which underlie cooperation, serve as a barometer for social cohesion. Low measures of trust elicit alarm, while high measures equate with societal vigor, e.g., "Trust enables cooperation, cooperation enables specialization, and specialization drives productivity" [2]. The role of cooperation and trust in a harmonious society thus receives considerable attention [3].

Human society also depends on energy. Since the first fires lit up cave dwellings, human societies have been using energy in many and various ways. Energy transitions (e.g., wood to coal, coal to petroleum) have had marked societal impacts as well. Currently, another energy transition is underway, from products of photosynthesis to more direct harvesting of abiotic energy sources (e.g., sun and wind). While history suggests that this will be a protracted affair [4], increasing concerns about greenhouse gas emissions lend an urgency that perhaps was absent from previous transitions. Energy will remain a preoccupation for policymakers for the foreseeable future.

Cooperation and energy have received considerable scientific attention as well. Indeed, in the mid- to late-twentieth century, large scientific conflicts flared in these two seemingly distinct fields of scientific inquiry. In bioenergetics, which examines how organisms obtain and utilize energy, the chemiosmotic hypothesis of Mitchell [5] suggested a novel mechanism for energy conversion. This hypothesis roiled the field and led to nearly two decades of the "oxphos wars," one of the longest, most contentious, and most consequential of all scientific debates. Common ground was eventually established [6], and Mitchell was awarded the Nobel Prize in 1978.

In evolutionary biology, meanwhile, Wynne Edwards [7] strongly articulated the view that organisms may act for the "good of the group." This work crystalized a long history of imprecise thinking about the evolution of cooperation. The resulting

N. W. Blackstone, *Energy and Evolutionary Conflict*,
https://doi.org/10.1007/978-3-031-06059-5_1

backlash led to near-absolute condemnation of any evolutionary model that relied on selection of groups of cooperating individuals, rather than on individuals furthering their own selfish replication [8, 9]. Even a cursory examination of the history of life, however, revealed that various biological units—individuals—have repeatedly banded together into groups (e.g., major transitions [10] of genes into chromosomes, prokaryotes into eukaryotes, and single cells into multicellular organisms). The juxtaposition of the seeming weakness of group-level selection on one hand, and the repeated emergence of groups in history of life on the other, stimulated considerable theoretical developments focused on cooperation and ultimately led to the emergence of the modern, multilevel theory of evolution [11].

While both controversies have received ample attention, no one has ever suggested that one might inform the other, i.e., that energy metabolism in general and chemiosmosis in particular might be relevant to the evolution of cooperation. For that matter, no one has ever suggested that there are mechanistic similarities underlying the evolution of cooperation in its various forms, e.g., symbiosis and multicellularity. To develop these connections, several introductory chapters will be devoted to providing background information on the seemingly disparate fields of energy metabolism, the evolution of cooperation, the history of life, and prior considerations of metabolism in organismal biology.

Perhaps a brief digression is also necessary in this context. Certainly, there have been a number of excellent studies pointing out the relationship between bioenergetics and complexity. As described in detail in Chap. 7, for instance, these studies clearly show that complex eukaryotic cells enjoy a number of energetic advantages over simple prokaryotic cells. Nevertheless, these advantages do not solve the problem of the evolutionary conflict inherent in forming the groups that became the cooperative cells. Evolution cannot work toward a fitness advantage that is not immediately conferred. At the start of any evolutionary transition, the higher-level units do not exist. Only lower-level units exist, and cooperation among these units entails evolutionary conflicts. It is to this problem that bioenergetic considerations are applied herein.

Indeed, the central idea being developed is remarkably simple and parallels a conundrum that confronts the drive to decarbonize human society. For instance, currently California leads the United States in "alternative" energy, obtaining roughly a third of its electricity from wind and solar. One unanticipated effect is that on windy, sunny days in southern California, the price of electricity drops as the power company tries to lure customers into using more electricity [12]. This seemingly counterproductive policy has a simple rationale: if demand does not match the increased supply, power lines may overheat and melt. Chemiosmosis may pose a similar dilemma for cells and organisms. As described in Chap. 2, chemiosmosis rapidly converts energy, and once storage capacity is exceeded, an overabundance of product has various negative consequences. In a sense, chemiosmosis behaves like a poorly insulated wire. When supply exceeds demand, the wire does not melt, but it will cast off electrons, which typically form potentially detrimental reactive oxygen species. While to some extent chemiosmotic processes can be modulated, under certain circumstances, it is also possible to simply disperse the products into the environment. In biological systems in which energy is frequently limiting, such largesse will not go unnoticed—the free lunch will attract a crowd. In other words, chemiosmosis can result in groups of individuals.

In considering the evolution of cooperation (Chap. 3), group formation is often a key step. Groups of individuals can facilitate kin selection and reciprocal altruism [13–15]. Groups also constitute a higher-level unit for selection to act on [16]. In groups of individuals, selection can under some circumstances favor cooperation rather than competition. Group formation, for whatever reason, can automatically potentiate the emergence of cooperation. Dispersing excess products of chemiosmosis into the environment for purely selfish reasons can thus be the first step to group formation and ultimately cooperation [17, 18]. A generation ago, sociobiology revolutionized the study of cooperation with a deceptively simple message—follow the genes [19]. Here, this message is complemented with an equally simple one—follow the electrons.

These ideas suggest a re-evaluation of the costs and benefits of cooperative relationships among biological units, e.g., mutualistic symbiosis and multicellularity. For instance, on sunny days, plants provide mycorrhizal fungi with abundant photosynthate. Is this a cost or a benefit? Similarly, on sunny days, corals with symbiotic algae release large amounts of nutrient-rich material into the ocean. Why would they do this? When defecting from the good of the multicellular group, mammalian cancer cells famously downregulate chemiosmosis when exhibiting the so-called Warburg effect. Again, why do they do this? Finally, the two most successful symbioses in the history of life, the mitochondrion and chloroplast, are both chemiosmotic. Coincidence? I think not.

With the history of life (Chap. 4), pre-chemiosmotic forays into the influence of metabolism (Chap. 5), and considerations of the effects of metabolism on group formation (Chap. 6) as further background, these themes will be more thoroughly developed in the remaining chapters. Chapter 7 examines how chemiosmosis can enlighten one of the great mysteries in the history of life, the origin of eukaryotes. Chapter 8 illuminates several modern symbioses in the same way, while Chaps. 9 and 10 examine another major evolutionary transition in this context, the origin of multicellularity. Further, if metabolism has influenced the history of life as suggested here, numerous vestiges or quirks should be found in modern biology. The ideas presented here provide a framework to rationalize at least some of these seemingly baroque natural histories as detailed in Chap. 11. Finally, some thoughts are provided as to why the fields of energy metabolism and the evolution of cooperation have remained distinct and how these fields can better come together to elucidate the metabolic roots of cooperation.

References

1. Lewis D (2020) Where COVID contract-tracing went wrong. Nature 588:384–387
2. Henderson MT (2019) How technology will revolutionize public trust. Wall Street J October 19–20, p C3
3. Newton K (2001) Trust, social capital, civil society, and democracy. Int Polit Sci Rev 22:201–214
4. Smil V (2010) Energy transitions: history, requirements, prospects. Praeger, Santa Barbara
5. Mitchell P (1961) Coupling of phosphorylation to electron and hydrogen transfer by a chemi-osmotic type of mechanism. Nature 191:144–148
6. Boyer PD, Chance B, Ernster L, Mitchell P, Racker E, Slater EC (1977) Oxidative phosphorylation and photophosphorylation. Annu Rev Biochem 46:955–1026
7. Wynne-Edwards VC (1962) Animal dispersion in relation to social behavior. Hafner, New York
8. Williams G (1966) Adaptation and natural selection. Princeton University Press, Princeton
9. Williams G (ed) (1971) Group selection. Aldine Atherton, Chicago
10. Maynard Smith J, Szathmáry E (1995) The major transitions in evolution. Oxford University Press, Oxford, pp 1–346
11. Radzvilavicius AL, Blackstone NW (2018) The evolution of individuality, revisited. Biol Rev 93:1620–1633
12. Service RF (2019) Renewable bonds. Science 365:1236–1239
13. Hamilton WD (1964) The genetical evolution of social behaviour. I. J Theor Biol 7:1–16
14. Hamilton WD (1964) The genetical evolution of social behaviour. II. J Theor Biol 7:17–52
15. Trivers RL (1971) The evolution of reciprocal altruism. Q Rev Biol 46:35–57
16. Wilson DS (1975) A theory of group selection. Proc Natl Acad Sci U S A 72:143–146
17. Blackstone NW (2020) Chemiosmosis, evolutionary conflict, and eukaryotic symbiosis. In: Kloc M (ed) Symbiosis: cellular, molecular, medical, and evolutionary aspects. Springer, Cham, pp 237–252
18. Blackstone NW, Gutterman JU (2021) Can natural selection and druggable targets synergize? Of nutrient scarcity, cancer, and the evolution of cooperation. BioEssays 43:2000160
19. Pinker S (2002) The blank slate. Penguin Books, New York

Chapter 2
Energy Conversion: How Life Makes a Living

In broadest terms, chemiosmosis is the movement of protons over a membrane (hence the resemblance in name to osmosis, the movement of water over a membrane). In respiration, what happens is this. Electrons are stripped from food and passed along a chain of carriers to oxygen. The energy released at several points is used to pump protons across a membrane. The outcome is a proton gradient over the membrane. The membrane acts a bit like a hydroelectric dam. Just as water flowing down from a hilltop reservoir drives a turbine to generate electricity, so in cells the flow of protons through protein turbines in the membrane drives the synthesis of ATP. This mechanism was totally unexpected: instead of having a nice straightforward reaction between two molecules, a strange gradient of protons is interpolated in the middle.

Nick Lane [1]

The emerging consensus on the effects of carbon dioxide on global climate [2] has focused increased attention on Earth's biogeochemical cycles in general and the carbon cycle in particular. Living things have in fact dominated biogeochemical cycles practically since the origin of life more than 3.5 billion years ago. The living mechanisms that have driven some of the most profound global changes are the molecular complexes that comprise electron transport chains and carry out a process of energy conversion called "chemiosmosis." Via respiration and photosynthesis, living things use chemiosmosis to convert environmental energy sources (e.g., substrate or light) into more useful forms. Chemiosmotic mechanisms are found in eukaryotic organelles such as mitochondria and chloroplasts as well as in most prokaryotes. The molecular complexes that carry out chemiosmotic processes, barely visible with an electron microscope, continue to profoundly influence biosphere- and geosphere-level processes. For instance, the energy flux produced by chemiosmosis considerably exceeds energy use by human society, and chemiosmotic mechanisms are central to producing the oxygen we breathe.

N. W. Blackstone, *Energy and Evolutionary Conflict*,
https://doi.org/10.1007/978-3-031-06059-5_2

In essence, energy conversion in virtually all living things depends on several well-matched, interacting parts [3–7]. First, cells or organelles require a membrane that is impermeable, particularly to charged ions. Second, embedded in this membrane must be metal-containing protein complexes that carry electrons. During the process of passing electrons from one to another, these electron carriers (collectively known as the electron transport chain) extrude protons. Thus, a "force field" of charged protons surrounds the membrane of the cell or organelle [6]. Finally, the membrane must contain protein complexes (known as ATP synthases) that function as "proton drains." When protons zip down these drains and return to the interior of the cell or organelle, they impart energy to a tiny mechanical mechanism that unites adenosine diphosphate (ADP) and inorganic phosphate to form adenosine triphosphate (ATP). ATP and related molecules function as the "battery" of virtually all modern living things.

The extent to which chemiosmosis was a unique and unexpected mechanism for energy conversion can only be understood by reviewing a bit of the history of the field of bioenergetics. Indeed, there are few better examples of the (sometimes messy) process of scientific progress. Referring to the nearly two decades of contention—the oxphos wars—surrounding the acceptance of the chemiosmotic mechanism, Harold [3] writes:

> A historical treatment of this upheaval would be both entertaining and instructive, not least as an illustration of the human quality of this supposedly detached and objective enterprise called science.

It is likely too soon for such a definitive history to be written, but aspects of this history continue to emerge and be debated [8–13]. This account has long antecedents, but the central issue can be succinctly summarized: while general ideas about metabolism and respiration have been part of scientific thinking for centuries, until relatively recently no hypothesis could clearly explain the central process of energy conversion. Respiration and the oxidation of carbon compounds were clearly linked to the synthesis of ATP, but a mechanistic explanation of the link remained elusive.

In the early 1960s, three competing hypotheses attempted to explain this mechanism: (1) chemical coupling, (2) conformational coupling, and (3) chemiosmosis. Chemical coupling was based on the long history of classical biochemistry. For instance, studies of glycolysis (Fig. 2.1) showed that ATP was generated by transfer of a phosphoryl group (PO_3^{-2}) from an organic compound to ADP. Adherents of hypothesis (1) postulated that some sort of similar intermediate trapped the energy from oxidation and triggered the formation of ATP. As described by Carafoli [13]: "The real challenge in the mitochondrial arena was, at the time, the mythical high-energy intermediate (colloquially called the 'squiggle')."

To some degree, hypothesis (2) was a reconceptualization of hypothesis (1) in which the energy from oxidation was captured by a high-energy conformation of a protein, which then synthesized ATP. As discussed below, conformational coupling has been found to play a central role in energy conversion. Peter Mitchell [15] proposed hypothesis (3), initially on the basis of very little experimental data. In broad outline, this is the chemiosmotic mechanism generally accepted today, although

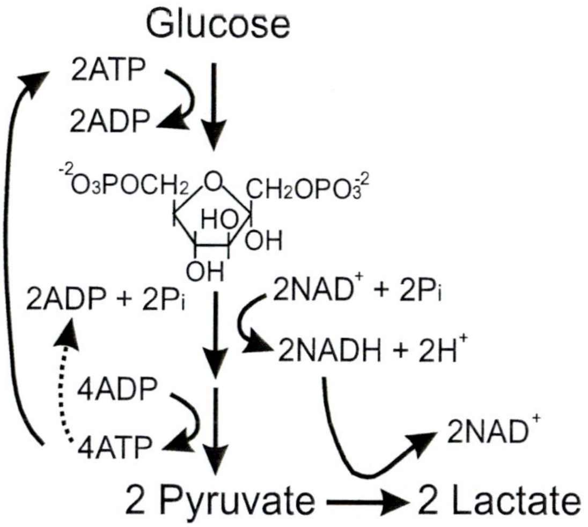

Fig. 2.1 An overview of glycolysis. Glucose is oxidized to pyruvate with the net gain of two molecules of ATP. Metabolism (dotted arrow) regenerates ADP and inorganic phosphate (P_i). In the absence of oxygen, NADH can be oxidized by conversion of pyruvate to lactate. In the presence of oxygen, pyruvate can be further oxidized in the Krebs cycle of mitochondria, with reducing equivalents powering the electron transport chain (Figs. 2.2 and 2.3). Glycolysis is an example of substrate-level phosphorylation in which ATP is generated by transfer of a phosphoryl group (PO_3^{-2}) from an organic compound to ADP. (Modified from Elliot and Elliot [14])

many, if not all, of the details differ from the original proposal [5, 6, 10]. In fact, Mitchell explicitly argued against many aspects of the modern view of chemiosmosis (e.g., proton pumping and conformational changes [10]). Honing the original hypothesis thus required a lengthy period of ferment, culminating with the widespread acceptance of the initial formulation of the modern chemiosmotic model [16] and the perhaps controversial awarding of the Nobel Prize for Chemistry to Peter Mitchell in 1978 [10]. Ultimately, the process exemplifies the collective power of the scientific method: testing and modifying a hypothesis by the scientific community until there is a good fit with the data.

The long delay from the initial proposal to its widespread acceptance resulted from the complexity of the experiments required to test the chemiosmotic hypothesis and the existence of complicating factors and alternative interpretations. As pointed out by Scheffler [5]:

> It is probably fair to say that, although Mitchell contributed to the experimental support of his hypothesis, his brilliant idea provoked many other talented individuals either to support or challenge it, and in the end, it was verified as another major landmark in the intellectual landscape of bioenergetics.

Leading scientists of the time used the predictions of the three alternative hypotheses to plan experiments and gather data [16]. For instance, Mitchell's hypothesis was based on a view of membrane biology that was only beginning to emerge at the time. Mitchell's hypothesis also required that electron transport chains extrude protons in one direction, while proton flow in the other direction was thought to drive ATP synthase. As well, Mitchell's hypothesis predicted characteristic relationships (i.e., stoichiometry) for the number of protons translocated per two electrons passing through the electron transport chain and for the number of protons required to make a single ATP from ADP and inorganic phosphate.

A wealth of experimental data gradually came to support chemiosmosis in broad outline [3, 5, 16]. Neurobiologists in particular showed that membranes frequently separate ion gradients. Various ion pumps that hydrolyzed ATP were identified. Running in reverse, these pumps could trigger ATP formation from ADP and inorganic phosphate. Bacterial models were developed, which because of their relative simplicity compared to mitochondria and chloroplasts allowed some hypotheses to be more easily tested. Various artificial membrane systems were also cleverly devised and utilized. With a few assumptions, predictions concerning the number of protons translocated per electron could yield predictable ATP to oxygen ratios. Such P/O ratios were measurable, and initial results tended to support chemiosmotic predictions. Nevertheless, because of a variety of "proton leaks" in membranes and "slip" in proton pumps [17], these ratios remained an area of contention for a number of years [18].

On the other hand, many aspects of the original view of chemiosmosis were ultimately falsified or at least heavily modified. Conformational changes in proteins were found to be an important and necessary aspect of chemiosmosis [16, 19–22]. Proton "pumping" of some sort seemed to be required in most cases for proton translocation. "Redox loops" turned out to be much less common than postulated by Mitchell. The notion of a redox loop is that a hydrogen atom (which consists of a single proton and a single electron) is first transported across the membrane, and then split into its component proton and electron. The proton contributes to the transmembrane electrochemical gradient, while the electron is transported back across the membrane. At least as originally proposed by Mitchell, the "Q cycle" was a sort of redox loop. The Q cycle refers to coenzyme Q (co-Q, $C_{59}H_{90}O_4$), sold commercially as the antioxidant "Co-Q10." Co-Q is a mobile electron carrier that shuttles electrons to complex III in mitochondrial electron transport chains (Fig. 2.2). Because co-Q is lipid soluble, it can in theory diffuse across the membrane, carrying hydrogen atoms with it. If co-Q can pass electrons to one of the membrane-bound complexes and at the same time release protons into the intermembrane space, some sort of proton-motive Q cycle is possible. In other words, co-Q could build a membrane potential merely by shuttling from one side of the membrane to another, picking up hydrogen atoms from one side and dropping off their component parts (protons and electrons) on the other. Later research, however, has shown that this view of the Q cycle vastly oversimplifies the reality of electron flow [12]. Even the modified form of the Q cycle that survives in most textbooks today may not accurately reflect this complexity [5, 24].

In addition to suggesting the complex, nonintuitive nature of chemiosmosis, this brief historical sketch illustrates several important points about science in general. Faced with several plausible hypotheses for energy conversion, scientists studying bioenergetics in the 1960s focused on developing experiments that could provide

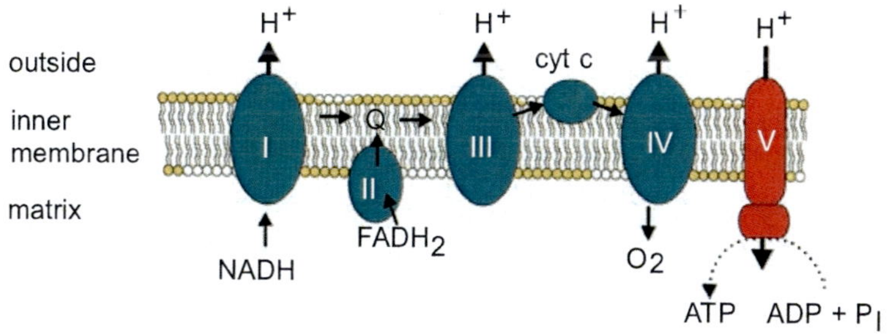

Fig. 2.2 Schemata of the mitochondrial electron transport chain, showing complexes I–V, coenzyme Q, and cytochrome *c*. Small arrows trace the flow of electrons from NADH and $FADH_2$ to oxygen. Large arrows show the extrusion of protons (H^+) by complexes I, III, and IV and the return of protons to the matrix via ATP synthase (complex V), triggering the assembly of ATP (dashed arrow). (From Blackstone [23])

data to support one of the alternatives and falsify the others. Modification of the original hypotheses continued apace with each new discovery. While the chemiosmotic hypothesis was ultimately accepted in broad outline, it was also recognized that the mechanisms of chemiosmosis in fact incorporate aspects of the other hypotheses. Even in the case of a scientific "revolution," the revolutionary theory ends up building on much of the previously existing knowledge.

In terms of the actual mechanisms of energy conversion, as Scheffler [5] emphasizes, the major insight of chemiosmosis was that:

> …a proton gradient across the inner membrane can be set up independently by electron transport (like water moved uphill into a dam by evaporation and condensation), and this gradient can be used to drive ATP synthesis (water running through turbines to produce electricity).

Building a proton gradient via electron transport is thus a common feature in all instances of chemiosmosis. Nevertheless, some of the specific components of chemiosmosis differ between bacteria, chloroplasts, and mitochondria. Since the emphasis of subsequent chapters is on mitochondria and chloroplasts, these mechanisms are the focus here (Fig. 2.2). Mitochondria, the powerhouses of eukaryotic cells, oxidize substrate (amino acids, carbohydrates, fatty acids) and reduce coenzymes NAD^+ (nicotinamide adenine dinucleotide) and FAD (flavin adenine dinucleotide). For instance, pyruvate from glycolysis can be further oxidized by the Krebs cycle, producing NADH and $FADH_2$ (Fig. 2.3). In turn, oxidation of NADH and $FADH_2$ provides electrons to the electron transport chain. Electron flow between the major complexes of this chain drives the extrusion of protons, establishing a steep electrochemical gradient across the inner mitochondrial membrane. This gradient powers most cellular functions, particularly by allowing the formation of ATP via ATP synthase [5, 6].

The major complexes of the electron transport chain (I–V in Fig. 2.2) are appropriately named, consisting of many different peptides and cofactors. In mammals, for instance, complex I consists of at least 42 peptides, while homologous

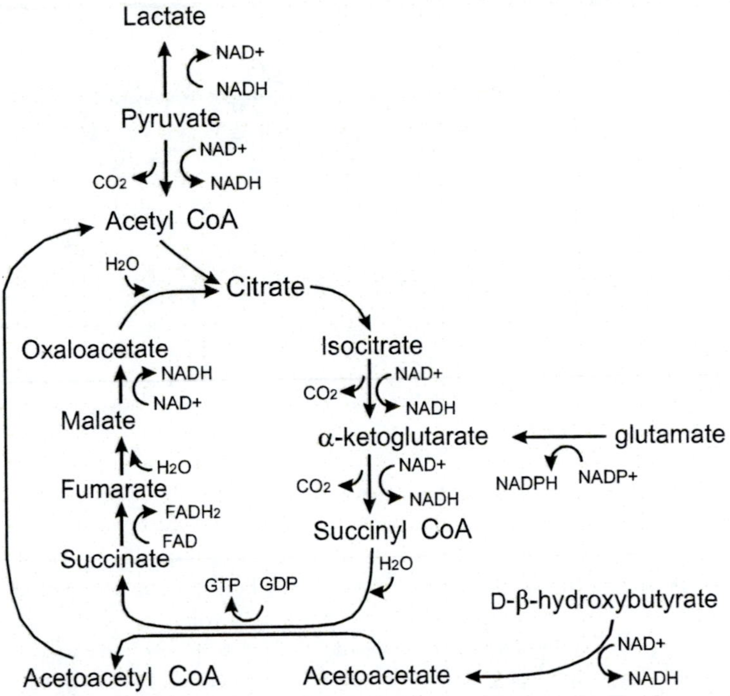

Fig. 2.3 Simplified diagram of the Krebs cycle and related metabolism. Pyruvate produced from glucose can be converted to lactate (Fig. 2.1) or broken down in mitochondria. The first step in this later process is catalyzed by pyruvate dehydrogenase and results in acetyl-CoA, which is fed into the Krebs cycle by reacting with oxaloacetate. In each turn of the cycle, the products include two molecules of CO_2, three molecules of NADH, one molecule of GTP, and one molecule of $FADH_2$. NADH and FADH2 are oxidized by the electron transport chain (Fig. 2.2) to build the transmembrane electrochemical gradient. (Modified from Sato et al. [25])

complexes in prokaryotes exhibit 14 peptides [5]. Electron flow within and between these peptides is facilitated by attached or embedded cofactors, some containing metal ions. In particular, iron (found in both heme and iron-sulfur clusters) plays a prominent role. As electrons flow between iron-containing redox centers, iron atoms are in turn reduced to ferrous ions (Fe^{2+}) then re-oxidized to ferric ions (Fe^{3+}). In contrast, the mobile electron carrier coenzyme-Q (co-Q) is a relatively simple, lipid-soluble molecule. As is typical of chemical antioxidants, co-Q is stable in several oxidation states. This allows co-Q to transport hydrogen atoms (each of which consist of a single proton and a single electron) through the membrane.

One of the crucial features of chemiosmosis was being elucidated even as Mitchell was first proposing his theory. Electron transfer within complexes of the electron transport chain showed remarkable insensitivity to temperature [26], suggesting some nonstandard mechanism. Sometimes referred to as "electron tunneling," such quantum electron transfer proceeds very rapidly, for instance, compared to soluble reactions [27]. In addition, later work showed the existence of "super-complexes" among the membrane-bound electron carriers [28]. Thus, electron transfer within and between membrane-bound complexes in chemiosmosis occurs extremely rapidly.

This rapidity poses problems in linking chemiosmosis to the soluble reactions that either feed reducing equivalents into electron transport chains or store the products that they produce. For instance, RuBisCO (ribulose-1,5-bisphosphate carboxylase/oxygenase), the enzyme that fixes CO_2 in photosynthesis, is perhaps the most abundant protein on Earth because it is "mopping up" the products of chemiosmosis in chloroplasts, ATP, and NADPH. The abundance of RuBisCO exemplifies how cells cope with the rapidity of chemiosmosis: deploying large quantities of the soluble enzymes that interact with electron transport chains.

By and large, this strategy works, and metabolic homeostasis can be maintained. Nevertheless, the linking of chemiosmosis to slower soluble reactions and potentially limited storage capacity has other consequences as well. If an accumulation of products inhibits electron flow, these electrons may divert to molecular oxygen and reactive oxygen species (i.e., partially reduced forms of oxygen, ROS) will form. By separating hydrogen atoms into protons and electrons, the chemiosmotic process itself is the cause of ROS formation. The role of ROS is discussed in more detail below.

The electron transport chain thus fits expectations that mechanisms of energy conversion should be among the most elaborate adaptations of living things. Hypotheses for how the electron transport chain evolved differ depending on whether the origin of life is thought to be heterotrophic or autotrophic. By the heterotrophic view (which in this case is meant to indicate metabolizing preexisting reduced carbon compounds), the first living things employed some sort of fermentative metabolism, similar to glycolysis (Fig. 2.1) [29]. The initial electron transport chain provided a mechanism to re-oxidize NADH while saving some pyruvate for biosynthesis [30]. A substrate-driven proton pump fulfilled this function [31]. Combined with another membrane-bound proton pump, in particular one that was ATP-driven, the two pumps would build a transmembrane proton gradient. At some point, the weaker pump (the ATP-driven one) began to run in reverse and became an ATP synthase. In one step, a functional electron transport chain was formed [29–31]. By one version of the autotrophic view, the proton gradient of the electron transport chain had an abiotic origin perhaps in microscopic fissures of rift vents, where alkaline fluid from the vent met acidic seawater. Life evolved by harnessing this gradient [1, 32, 33]. Whether the origins of the electron transport chain were heterotrophic or autotrophic, once a rudimentary chain existed, natural selection would begin to hone this new adaptation. The power of natural selection in this context cannot be doubted, as it is routinely used as a tool in biotechnological applications that are related to energy conversion [34]. A prokaryote level of complexity for energy conversion would likely be rapidly attained.

While of course chloroplasts require light, in many ways, they function in a similar manner as mitochondria. Electrons (from water in the former or coenzymes such

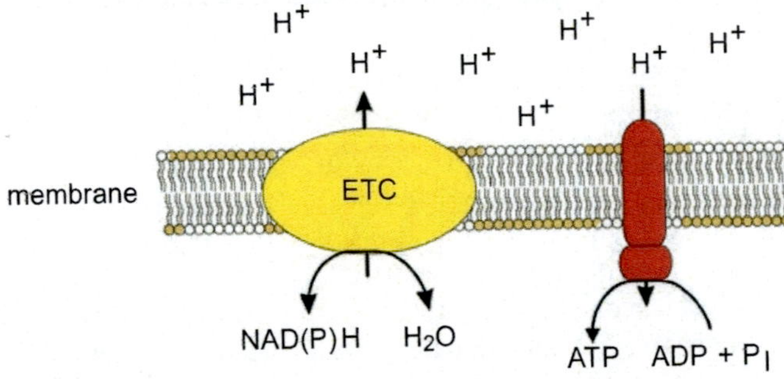

Fig. 2.4 Schematic summarizing eukaryotic chemiosmosis. Mitochondria oxidize reduced cofactors such as NADH, run the electrons through an electron transport chain (ETC), and build a transmembrane proton gradient. Protons return via ATP synthase, triggering the formation of ATP from ADP and inorganic phosphate (Pi). With the input of light energy, chloroplasts oxidize water, run the electrons through an electron transport chain (which is homologous to that in mitochondria), and build a transmembrane proton gradient. As in mitochondria, protons trigger the formation of ATP and electrons reduce NADP+ to NADPH. Mitochondria can store ATP as phosphoenolpyruvate, phosphocreatine, or similar compounds, while chloroplasts store the energy in ATP and NADPH by fixing carbon via the soluble enzyme RuBisCO. (From Blackstone [35])

as NADH in the latter) power an electron transport chain, producing a proton gradient, which catalyzes the formation of ATP in both chloroplasts and mitochondria (NADPH is also formed in the former) (Fig. 2.4). Cells containing mitochondria can then store ATP as phosphoenolpyruvate or phosphocreatine or something similar, while chloroplasts store the energy in ATP and NADPH by fixing carbon via the soluble enzyme RuBisCO.

To recapitulate, chemiosmotic mechanisms are composed of several parts: (1) a membrane that is impermeable to protons and other charged ions, (2) metal-containing protein complexes that pump or extrude protons in response to electron flow, and (3) ATP synthase through which protons return to the mitochondrial matrix and in the process trigger the formation of ATP from ADP and phosphate. The interaction of these parts contributes to the basic function. The proton-pumping complexes translocate protons across the membrane. The membrane itself blocks the return of these protons except via specified channels. ATP synthase channels these protons into triggering the formation of ATP. As evidenced by human mitochondrial diseases [5], weakening or removing any one of these component parts causes the system to effectively cease functioning. Pharmacological manipulations can also demonstrate the crucial nature of each component. For instance, chemical "uncouplers" provide an alternative channel through the membrane so that oxidation of substrate continues, often at a very rapid pace, but no ATP is synthesized [17]. Because the extruded protons are not effectively blocked by the membrane, they can return to the mitochondrial interior without triggering ATP formation. Various substances that inhibit electron flow between protein complexes or to the terminal electron acceptor, oxygen, also halt the process of energy conversion. We die when we breathe carbon monoxide (or cyanide gas) because these substances bind to complex IV and prevent electrons from reaching oxygen [36]. Finally, inhib iting ATP synthase (for instance, with oligomycin) also blocks the chemiosmotic process.

These sorts of pharmacological manipulations not only provide specific insight into the functions of parts of the electron transport chain but also illuminate the crucial workings of the electron transport chain in its entirety. For instance, uncouplers were once used to trigger human weight loss because of their stimulation of metabolism. Why should a substance that allows protons to pass through the inner mitochondrial membrane also result in weight loss in human beings? The answer illuminates one of the oldest functions of the electron transport chain as a whole: it serves as a mechanism that living things use to sense their environment. This environmental sensing in turn triggers a variety of cellular and organismal responses even in modern organisms.

To see how this is accomplished, consider that the electron transport chain links what are ultimately environmental sources and sinks of electrons (e.g., food and oxygen). Food is oxidized and coenzymes (e.g., NAD^+ and FAD) are reduced. These coenzymes are then oxidized by components of the electron transport chain which in turn become reduced. As electrons are passed between complexes, this process of oxidation and reduction continues. Finally, the electrons are deposited on diatomic oxygen, which is reduced to water. Bioenergetic metabolism thus links external electron sources and sinks through a series of living redox couples [37]. Ultimately, via the process of oxidation and reduction, the organism obtains energy from the environment. This has been a common theme of life since before the last common ancestor of all life, and it highlights the crucial mechanism that organisms use to detect features of the environment. For instance, if the coenzymes are oxidized (e.g., NAD^+ and FAD), the organism is running out of food and the ATP/ADP ratio approaches zero. Successful organisms have quickly responded to such signals for thousands of millions of years (e.g., EAT!). On the other hand, if most of the coenzymes remain reduced (e.g., NADH and $FADH_2$), various other responses may be indicated. The number of electron transport chains may be insufficient for the available food and transcription and translation may ensue. Alternatively, the terminal electron acceptor (e.g., oxygen) might be scarce. Several organismal responses might ensue, involving behavior and locomotion (e.g., MOVE!), physiology (e.g., BREATH!), or development (e.g., angiogenesis). In this context, the response of human beings to uncouplers can be seen as a desperate physiological battle to convert ADP into ATP, even though the mechanism to do this (the proton gradient) has been abolished.

In these and many other circumstances, living things employ this sort of "redox" signaling as a rapid and effective mechanism to adjust bioenergetic metabolism to environmental conditions. Throughout the history of life, the "payoff" has been a larger dividend of energy. Since this dividend can be spent on a higher rate of replication, the Darwinian imperative is clear. Insight into how such redox signaling works can be gained by examining modern bacteria. While bacteria may constitute the simplest extant cells, they are nevertheless far from simple. Redox signaling is among the many sophisticated mechanisms that bacteria can deploy. Control of gene expression commonly occurs in response to changes in redox potential, which are ultimately linked to environmental factors (e.g., light, oxygen, or particular types of food). Two-component signal-transduction systems are often used in adapting bacterial metabolism to environmental conditions [37, 38]. The Arc system of *Escherichia coli* provides a well-studied example. Two proteins, ArcA and ArcB, are involved. ArcB is a transmembrane sensor kinase with a loop exposed to the cytoplasm. The cytoplasmic loop contains a conserved histidine residue, which is the site of autophosphorylation. This occurs in response to the redox state of quinone electron carriers of the electron transport chain (comparable to CoQ in Fig. 2.2; note that the electron transport chain of *E. coli* is similar to, but not entirely the same as, the one illustrated). Oxidized forms of quinone inhibit autophosphorylation [38]; if quinones are reduced, this inhibition is removed. Subsequent to autophosphorylation, ArcB transphosphorylates the second component, ArcA, which is a global transcriptional regulator. When phosphorylated, ArcA represses the expression of many genes whose products are involved in aerobic respiration and activates many of the genes whose products are involved in anaerobic fermentation.

The logic of the ArcA-ArcB system is thus readily apparent [37]. The quinone electron carriers remain relatively oxidized, and autophosphorylation is inhibited as long as electron transport to the terminal electron acceptor (diatomic oxygen) is possible. If oxygen is not available, electrons "back up" on the electron carriers of the electron transport chain, and these carriers become reduced. Autophosphorylation and transphosphorylation ensue, and the components of anaerobic metabolism are activated. Many aspects of bacterial responses to environmental signals are mediated in comparable ways. No doubt such mechanisms are in part responsible for the success that bacteria have enjoyed throughout the history of life.

In eukaryotes, redox signaling is often accomplished by intermediaries known as reactive oxygen species (ROS, here broadly defined as partially reduced forms of diatomic oxygen such as superoxide and hydrogen peroxide). For instance, if the environment becomes stressful and thus unfavorable for cellular replication, yet plenty of food remains available, cellular metabolic demand will diminish, and the ATP/ADP ratio will approach one. Oxidation of food, however, will continue until the electrochemical gradient is maximal and the membrane-bound electron carriers are highly reduced. In such circumstances, these electron carriers will typically donate electrons to molecules (e.g., diatomic oxygen) whose partially reduced products can serve as messengers (e.g., ROS) [39–42]. Such messengers can then trigger the appropriate adaptive response at the cellular level [43–48]. On the other hand, during starvation the ATP/ADP ratio will approach zero, and the electrochemical gradient becomes minimal. The electron carriers now are relatively oxidized and formation of, for instance, ROS is also minimal. Again, appropriate responses can be subsequently triggered.

Much of the early research on ROS focused on the toxic effects of high concentrations of these molecules [49]. Nevertheless, it has become increasingly clear that ROS also play an important role in within- and between-cell signaling [39–42, 50]. Indeed, the risks of the toxic effects of hydrogen peroxide on one hand and the benefits of its signaling functions on the other are nicely summed up by the structure and function of peroxiredoxin proteins [51]. When H_2O_2 is reduced, the catalytic cysteine residue of the peroxiredoxin is oxidized. At low concentrations of peroxide, a disulfide bond will form with the resolving cysteine from the other subunit of the homodimer. Thioredoxin subsequently reduces the disulfide bond, and the enzyme is once again ready to interact with another peroxide molecule. On the other hand, at high concentrations of peroxide such as those typically found in a signaling pulse, there is a high probability that the oxidized cysteine will react with another H_2O_2 molecule before it reacts with the resolving cysteine. The higher oxidation products thus formed cannot be reduced by thioredoxins [52, 53]. The antioxidant function of the peroxiredoxin is then inactivated. The peroxide that is spared can subsequently function as a messenger in signaling pathways, such as those activated by protein tyrosine kinases. For instance, protein tyrosine phosphatase (PTP) 1B antagonizes the effects of these kinases, but PTP can be inactivated when the catalytic cysteine is oxidized by H_2O_2. Thus, the activated pathway can proceed. These pathways, as well as peroxiredoxins and thioredoxins, seem to be evolutionarily ancient [54, 55].

It would be misleading, however, to imply that ROS are always benign. As reported by Brownlee [56]:

> Diabetes-specific microvascular disease is a leading cause of blindness, renal failure and nerve damage, and diabetes-accelerated atherosclerosis leads to increased risk of myocardial infarction, stroke and limb amputation. Four main molecular mechanisms have been implicated in glucose-mediated vascular damage. All seem to reflect a single hyperglycaemia-induced process of overproduction of superoxide by the mitochondrial electron-transport chain. This integrating paradigm provides a new conceptual framework for future research and drug discovery.

Thus, a major disease of modern human society has its roots in microscopic organelles that misinterpret environmental signals and release ROS to the detriment of the human organism. Mitochondria are of course remnants of bacteria that became symbionts within the eukaryotic cells. In human beings, trillions of these eukaryotic cells now make up a single organism. For these three levels of the biological hierarchy to function properly, multilevel cooperation must occur. Yet in the case of diabetes, cooperation is elusive—the mitochondria are releasing toxic by-products of chemiosmotic metabolism to the detriment of the cell and the organism. As suggested in later chapters, it may not be a coincidence that this occurs under conditions of nutrient abundance. Before returning to this topic, however, more must be said about the evolution of cooperation among other topics.

References

1. Lane N (2009) Life ascending: the ten great inventions of evolution. Oxford University Press, Oxford
2. Walker G (2006) The tipping point of the iceberg. Nature 441:802–805
3. Harold FM (1986) The vital force: a study of bioenergetics. WH Freeman, New York
4. Saraste M (1999) Oxidative phosphorylation at the fin de siècle. Science 283:1488–1493

5. Scheffler IE (1999) Mitochondria. John Wiley, New York
6. Lane N (2005) Power, sex, suicide: mitochondria and the meaning of life. Oxford University Press, Oxford
7. Scheffler IE (2008) Mitochondria, 2nd edn. John Wiley, New York
8. Weber BH (1991) Glynn and the conceptual development of the chemiosmotic theory: a retrospective and prospective view. Biosci Rep 11:577–617
9. Williams RJP (1993) The history of proton-driven ATP formation. Biosci Rep 13:191–212
10. Malmström BG (2000) Mitchell saw the vista, if not the details. Nature 403:356
11. Harold FM (2001) Gleanings of a chemiosmotic eye. BioEssays 23:848–855
12. Crofts AR (2004) The Q-cycle—a personal perspective. Photosynth Res 80:223–243
13. Carafoli E (2003) Historical review: mitochondria and calcium: ups and downs of an unusual relationship. Trends Biochem Sci 28:175–181
14. Elliot WH, Elliot DC (1997) Biochemistry and molecular biology. Oxford University Press, Oxford
15. Mitchell P (1961) Coupling of phosphorylation to electron and hydrogen transfer by a chemiosmotic type of mechanism. Nature 191:144–148
16. Boyer PD, Chance B, Ernster L, Mitchell P, Racker E, Slater EC (1977) Oxidative phosphorylation and photophosphorylation. Annu Rev Biochem 46:955–1026
17. Kadenbach B (2003) Instrisic and extrinsic uncoupling of oxidative phosphorylation. Biochim Biophys Acta 1604:77–94
18. Hinkle PC (2005) P/O ratios of mitochondrial oxidative phosphorylation. Biochim Biophys Acta 1706:1–11
19. Efremov RG, Baradaran R, Sazanov LA (2010) The architecture of respiratory complex I. Nature 465:441–445
20. Efremov RG, Sazanov LA (2011) Structure of the membrane domain of respiratory complex I. Nature 476:414–420
21. Hunte C, Zickermann V, Brandt U (2010) Functional modules and structural basis of conformational coupling in mitochondrial complex I. Science 329:448–451
22. Sazanov LA (2015) A giant molecular proton pump: structure and mechanism of respiratory complex I. Nat Rev Mol Cell Biol 16:375–388
23. Blackstone NW (2003) Redox signaling in the growth and development of colonial hydroids. J Exp Biol 206:651–658
24. Osyczka A, Moser CC, Daldal F, Dutton PL (2004) Reversible redox energy coupling in electron transfer chains. Nature 427:607–612
25. Sato K, Kashiwaya Y, Keon CA, Tsuchiya N, King MT, Radda GK, Chance B, Clarke K, Veech RL (1995) Insulin, ketone bodies, and mitochondrial energy transduction. FASEB J 9:651–658
26. Chance B, Nishimura M (1960) On the mechanism of chlorophyll-cytochrome interaction: the temperature insensitivity of light-induced cytochrome oxidation in chromatium. Proc Natl Acad Sci U S A 46:19–24
27. Moser CC, Keske JM, Warncke K, Farid RS, Dutton PL (1992) Nature of biological electron transfer. Nature 355:796–802
28. Dudkina NV, Eubel H, Keegstra W, Boekema EJ, Braun H-P (2005) Structure of a mitochondrial supercomplex formed by respiratory-chain complexes I and III. Proc Natl Acad Sci U S A 102:3225–3229
29. Wilson TH, Lin ECC (1980) Evolution of membrane bioenergetics. J Supramol Struct 13:421–446
30. Gest H (1980) The evolution of biological energy-transducing systems. FEMS Microbiol Lett 7:73–77
31. de Duve C (1995) Vital dust. Basic Books, New York
32. Lane N, Allen JF, Martin W (2010) How did LUCA make a living? Chemiosmosis in the origin of life. BioEssays 32:271–280
33. Allen JF (2010) Redox homeostasis in the emergence of life. On the constant internal environment of nascent living cells. J Cosmol 10:3362–3373

34. Wang Y, Manow R, Finan C, Wang J, Garza E, Zhou S (2010) Adaptive evolution of nontransgenic *Escherichia coli* KC01 for improved ethanol tolerance and homoethanol fermentation from xylose. J Ind Microbiol Biotechnol. https://doi.org/10.1007/s10295-010-0920-5
35. Blackstone NW (2020) Chemiosmosis, evolutionary conflict, and eukaryotic Symbiosis. In: Kloc M (ed) Symbiosis: cellular, molecular, medical, and evolutionary aspects. Springer, Cham, pp 237–252
36. Pearce LL, Bominaar EL, Hill BC, Peterson J (2003) Reversal of cyanide inhibition of cytochrome c oxidase by the auxiliary substrate nitric oxide. J Biol Chem 278:52139–52145
37. Allen JF (1993) Control of gene expression by redox potential and the requirement for chloroplast and mitochondrial genomes. J Theor Biol 165:609–631
38. Georgellis D, Kwon O, Lin ECC (2001) Quinones as the redox signal for the arc two-component system of bacteria. Science 292:2314–2316
39. Armstrong JS, Whiteman M, Yang H, Jones DP (2004) The redox regulation of intermediary metabolism by a superoxide-aconitase rheostat. BioEssays 26:895–900
40. Salmeen A, Andersen JN, Meyers MP, Meng T-C, Hinks JA, Tonks NK, Barford D (2003) Redox regulation of protein tyrosine phosphatase 1B involves a suphenyl-amide intermediate. Nature 423:769–773
41. van Montfort RLM, Congreve M, Tisi D, Carr R, Jhoti H (2003) Oxidation state of the active-site cysteine in protein tyrosine phosphatase 1B. Nature 423:773–777
42. Filomeni G, Rotilio G, Ciriolo MR (2005) Disulfide relays and phosphorylation cascades: partners in redox-mediated signaling pathways. Cell Death Differ 12:1555–1563
43. Coloff JL, Rathmell JC (2006) Metabolic regulation of Akt: roles reversed. J Cell Biol 175:945–947
44. Kondoh H, Lleonart ME, Bernard D, Gil J (2007) Protection from oxidative stress by enhanced glycolysis: a possible mechanism of cellular immortalization. Histol Histopathol 22:85–90
45. Ladurner AG (2006) Rheostat control of gene expression by metabolites. Mol Cell 24:1–11
46. Coffman JA, Davidson EH (2001) Oral-aboral axis specification in the sea urchin embryo. Dev Biol 230:18–28
47. Coffman JA, McCarthy JJ, Dickey-Sims C, Robertson AJ (2004) Oral-aboral axis specification in the sea urchin embryo II. Mitochondrial distribution and redox state contribute to establishing polarity in *Strongylocentrotus purpuratus*. Dev Biol 273:160–171
48. Fomenko DE, Xing W, Adair BM, Thomas DJ, Gladyshev VN (2007) High-throughput identification of catalytic redox-active cysteine residues. Science 315:387–389
49. Gilbert DL (2000) Fifty years of radical ideas. In: Chiueh CC (ed) Reactive oxygen species: from radiation to molecular biology, Annals of the New York Academy of Sciences, vol 899. New York Academy of Sciences, New York, pp 1–14
50. Finkel T (2001) Reactive oxygen species and signal transduction. IUBMB Life 52:3–6
51. Georgiou G, Masip L (2003) An overoxidation journey with a return ticket. Science 300:592–594
52. Wood ZA, Poole LB, Karplus PA (2003) Peroxiredoxin evolution and the regulation of hydrogen peroxide signaling. Science 300:650–653
53. Jönsson TJ, Johnson LC, Lowther WT (2008) Structure of the sulphiredoxin-peroxiredoxin complex reveals an essential repair embrace. Nature 451:98–101
54. Chan TA, Chu CA, Rauen KA, Kroiher M, Tatarewicz SM, Steele RE (1994) Identification of a gene encoding a novel protein-tyrosine kinase containing SH2 domains and ankyrin-like repeats. Oncogene 9:1253–1259
55. Blackstone NW, Cherry KS, Van Winkle DH (2004) The role of polyp-stolon junctions in the redox signaling of colonial hydroids. Hydrobiologia 530(531):291–298
56. Brownlee M (2001) Biochemistry and molecular cell biology of diabetic complications. Nature 414:813–819

Chapter 3
The Puzzle of Cooperation

Next, the view of evolution as chronic bloody competition among individuals and species, a popular distortion of Darwin's notion of 'survival of the fittest,' dissolves before a new view of continual cooperation, strong interaction, and mutual dependence among life forms. Life did not take over the globe by combat, but by networking. Life forms multiplied and complexified by co-opting others, not by killing them.

Lynn Margulis and Dorion Sagan [1]

In 1859, Darwin published the *On the Origin of Species*, in which he outlined his theory of evolution [2]. He perhaps most succinctly articulated his theory in the introductory lines of a later book, the *The Descent of Man*, first published in 1871 [3]:

> He who wishes to decide whether man is the modified descendent of some pre-existing form, would probably first enquire whether man varies, however slightly, in bodily structure and in mental faculties; and if so, whether variations are transmitted to his offspring in accordance with the laws which prevail with the lower animals... The enquirer would next come to the important point, whether man tends to increase at so rapid a rate, as to lead to occasional severe struggles for existence; and consequently to beneficial variants, whether in body or mind, being preserved, and injurious ones eliminated.

In short, according to Darwin, evolution occurs when heritable variation is subject to natural selection.

In the *Origin of Species*, Darwin's stated goal was to demonstrate that the theory of "special creation"—that each species was created separately by God—was unnecessary. In this regard, he largely succeeded. Nevertheless, in his lifetime, his theory of evolution by natural selection had few adherents because of nagging questions: Where did variation come from? How was it inherited? In later editions of the *Origin*, Darwin invented increasingly fanciful answers to these questions. It was not until Mendel's work was rediscovered in the early twentieth century that one class of mechanisms of variation and inheritance began to be elucidated by the science of

N. W. Blackstone, *Energy and Evolutionary Conflict*,
https://doi.org/10.1007/978-3-031-06059-5_3

genetics. In time, the modern synthesis of the 1930s and 1940s unified Darwin's theory and Mendelian genetics [4].

Nevertheless, not all aspects of the natural world could be easily integrated into the modern synthesis. Cooperation was foremost among these unresolved issues: how can cooperation evolve if Darwinian competition is the driving forces of evolution? In fact, Darwin provided one answer to this puzzle, also in the *The Descent of Man* [3]:

> It must not be forgotten that although a high standard of morality gives but a slight or no advantage to each individual man and his children over the other men of the same tribe, yet that an increase in the number of well-endowed men and an advancement in the standard of morality will certainly give an immense advantage to one tribe over another. A tribe including many members who, from possessing in a high degree the spirit of patriotism, fidelity, obedience, courage, and sympathy, were always ready to aid one another, and to sacrifice themselves for the common good, would be victorious over most other tribes; and this would be natural selection. At all times throughout the world tribes have supplanted other tribes; and as morality is one important element in their success, the standard of morality and the number of well-endowed men will thus everywhere tend to rise and increase.

In this passage, Darwin focuses on a trait—morality—that is assumed to be inherited at least in part (e.g., Pinker [5]) and that "… gives but a slight or no advantage…" at the level of the human individual. In other words, at this level of the biological hierarchy, morality is selectively neutral. When individual-level selection alone operates, moral individuals will on average have no more offspring than immoral ones. Thus, the frequency of moral individuals will neither increase nor decrease. Darwin then points out that at a higher biological level—the tribe—the results of selection are quite different: "A tribe including many members who, from possessing in a high degree the spirit of patriotism, fidelity, obedience, courage, and sympathy…would be victorious over most other tribes…." In other words, when between-tribe conflict occurs, tribes that contain many moral individuals will prevail over tribes with fewer such individuals. Tribes that in aggregate have a high moral standard will increase in frequency relative to tribes that in aggregate have a low moral standard. The effects of tribe-level selection thus differ from the effects of individual-level selection. The latter will not affect the frequency of individuals that vary in moral standard, while the former very clearly does affect the frequency of tribes that in aggregate vary in moral standard. If between-tribe selection was a potent force in human evolution, the existence of human morality can be explained by this sort of natural selection [6]. Darwin's example encapsulates what later became known as "group selection": in a structured population, selection favors different traits at the individual and group level, and if between-group selection outweighs within-group selection, the trait favored at the group level can come to predominate.

By and large, Darwin's insight was lost by the early twentieth century as the modern synthesis began to be formulated. Advocates of cooperation thus turned toward other ideas. Early-twentieth-century formulations of the endosymbiont theory of the origin of eukaryotes explicitly rejected "Darwinian" notions of conflict and posed cooperation as an alternative. Such views were common particularly in

Russia [7]. For example, Mereschkowsky [8, 9] makes no mention of conflict (but see Sapp et al. [10]). Later, recognizing that the endosymbiont theory of the origin of mitochondria involved cooperation, Wallin [11] described "symbionticism" as a missing part of Darwin's theory, seemingly outside the realm of natural selection: "Modern writers have recognized the insufficiency of Darwin's hypothesis to explain the origin of species. The 'unknown factor' in organic evolution has been especially emphasized by Osborne, Bateson, Kellog, and other recent writers. This "unknown factor" is especially concerned with the origin of species." According to Wallin, the unknown factor was what he termed symbionticism. This predisposition against what was an oversimplified view of Darwinian evolution persisted in discussions of the endosymbiont theory for most of the twentieth century. For instance, when resurrecting this theory, Margulis [12] had little to say about potential conflicts, despite discussing scenarios in which evolutionary conflict would seem inevitable. Rather, as suggested in the epigraph, competition was somehow deemed illusory with cooperation prevailing, perhaps because it harmonized with social themes of the 1960s and 1970s.

Nevertheless, to some extent, the modern synthesis did include the recognition of circumstances not entirely unlike those mentioned by Darwin's example of human morality. For instance, Wright [13] and particularly in later works highlighted structured populations of many small groups with little between-group migration and potentially high levels of genetic drift. These circumstances could in theory provide fertile ground for the evolution of cooperation as later workers would point out. Many contemporaries, however, deemed these conditions unlikely in natural populations.

For a lengthy period, mainstream evolutionary biology viewed cooperation as of limited interest [14]. Eventually, however, the topic was taken up in earnest with a focus on natural histories entirely different from symbiosis. It would perhaps not be an exaggeration to suggest that these considerations of the evolution of cooperation have a long and contentious history [15–17], and relevant aspects of this history are briefly reviewed here. Some early discussions of selection used sloppy language, e.g., presenting selection as a force that acts for the benefit of the species. Potential misunderstandings became considerably more explicit with the work of Wynne-Edwards [18], who famously argued that species could be selected to limit their population sizes so as not to overexploit their food resources. A number of evolutionary biologists, most notably Maynard Smith [19] and Williams [20], pointed out the obvious limitations of this argument: if some individuals reproduce indiscriminately, while the rest of the population limits its reproduction, the prolific individuals will leave more offspring. If this behavior is inherited, at least in part, it will increase in frequency in the population, even if this leads to the destruction of habitat and the eventual demise of the species. In many circumstances, the high rate of selection at the level of the individual can thus overpower the slower rate of selection at the level of the group or species. Maynard Smith pointed out that the circumstances required for this scenario not to occur were those highlighted earlier by Wright: a population structured into many small groups in which genetic drift could be a potent factor [19]. Like many of Wright's contemporaries, Maynard Smith considered these conditions to be unlikely in natural populations.

These discussions led to broader considerations of the circumstances that may sometimes favor cooperation. Hamilton [21] introduces the subject:

> With very few exceptions, the only parts of the theory of natural selection which have been supported by mathematical models admit no possibility of the evolution of any characters which are on average to the disadvantage of the individuals possessing them. If natural selection followed the classical models exclusively, species would not show any behaviour more positively social than the coming together of the sexes and parental care.

Building on suggestions made by Haldane [22], Hamilton elaborates the theory of inclusive fitness (also known as kin selection), which had an enormous impact on subsequent evolutionary thinking [23]. Like all remarkable ideas, its basis is rather simple: to the extent that cooperation is genetically determined, individuals should cooperate to the degree that they are likely to share the same alleles. Thus, cooperation among full siblings should exceed cooperation among cousins, and so on. As summarized by Hamilton [24]:

> In brief outline, the theory points out that for a gene to receive positive selection it is not necessarily enough that it should increase the fitness of its bearer above the average if this tends to be done at the heavy expense of related individuals, because relatives, on account of their common ancestry, tend to carry replicas of the same gene; and conversely that a gene may receive positive selection even though disadvantageous to its bearers if it causes them to confer sufficiently large advantages on relatives.

In the context of these ideas, Hamilton [24] discusses numerous natural history examples. Many of the examples focused on group-living organisms and the actions toward others in their group: "The social behaviour of a species evolves in such a way that in each distinct behaviour-evoking situation the individual will seem to value his neighbours' fitness against his own according to the coefficients of relationship appropriate to that situation." The examples that perhaps captured the imagination of most readers were those involving the social insects, and the role of haplodiploid sex determination in boosting the coefficient of relatedness among workers in colonies of these insects. These examples, while a stunning example of the power of the inclusive-fitness approach, ultimately remain unsupported by the available data [4].

Hamilton's theory of inclusive fitness clearly had a major influence on Trivers' development of the idea of reciprocal altruism. Trivers [25] begins by elaborating the example of a drowning man and an unrelated rescuer:

> Were this an isolated event, it is clear that the rescuer should not bother to save the drowning man. But if the drowning man reciprocates at some future time, and if the survival chances are then exactly reversed, it will have been to the benefit of each participant to have risked his life for the other.

As Trivers points out, the circumstances that favor reciprocal altruism include group-living organisms with perhaps well-developed cognition: "Because they also meet the other conditions outlined here, primates are almost ideal species in which to search for reciprocal altruism." He does not rule out, however, a broader role for reciprocity, for example: "Reciprocal altruism can also be viewed as a symbiosis, each partner helping the other while he helps himself. The symbiosis has a time lag, however; one partner helps the other and must then wait a period of time before he is helped in turn." While reciprocal altruism would seem applicable to group-living primates, at a simpler level, syntrophic groups of microbes may engage in trading metabolic favors, again with a cooperative outcome [26].

Trivers [25] also introduced the idea of viewing cooperation as a solution to the "prisoner's dilemma," in which two prisoners can either defect (i.e., "rat out" the other) or cooperate and "keep mum." Their prison sentence depends on their choice of action and the choice of their accomplice. (Note that cooperation and defection are relative to each other, not relative to the "authorities"). In many cases, defection is favored, but under some circumstances, cooperation can emerge.

Reciprocity of some sort likely underlies human economic games [27]. "Public goods" games are widely examined. In the ultimatum game, for instance, one player offers to split some amount of money, and the second player can accept the split or reject it, in which case neither player gets any money. These games generally show that simple economic calculations fail to explain human behavior, and considerations of "fairness" are paramount. Economic theory suggests one should accept any amount >0 while offering much less than 50%. Nevertheless, offers are usually higher and rejections much more frequent than predicted by these simple considerations. We can also see this issue of fairness as central to many national and local debates: who is or is not contributing their "fair share?" Perhaps too often, everyone agrees that in principle a fair share should be contributed, but nobody agrees on exactly what that share actually should be!

Both Hamilton and Trivers examine cooperation and evolutionary conflict from a within-group perspective. While their ideas apply most clearly to group-living organisms (e.g., groups of kin or groups of reciprocators), their focus is entirely on the effects of selection on individuals, with their examples largely drawn from the natural histories of multicellular organisms. Their perspective thus differs from the group-selection examples developed by Darwin, Wright, Maynard Smith, and others.

Wilson [28] also developed a group-selection model, in which between-group selection overcomes within-group selection under some circumstances. While Wilson's model invokes a structured population, in contrast to models developed by Wright, Maynard Smith, and others, it does not rely on limited dispersal or genetic drift. Thus, in a population structured into "trait groups," two types of individuals exist, cooperators and defectors. An individual's reproductive success depends on the group that it is in. While having cooperators in a group increases average reproductive success, in mixed groups, defectors always have more offspring. Cooperators can be thought of as producing some kind of "public good," and it costs them to do this, but the group benefits. Hence, the frequency of cooperators declines in mixed groups. Under certain assumptions, however, when individuals mingle to form new groups, the frequency of cooperators has increased. Between-group selection outweighs within-group selection and cooperation spreads. As Wilson points out:

> Notice that this form of group selection never really violates the concept of individual selection. It is always the type with the highest per capita fitness that is chosen, but when the effect of more than one trait-group is considered, these are the very types that behave altruistically.

Indeed, depending on how this example is viewed, it may appear not to be group selection at all [4]. Further, while Wilson's goal was to develop a more broadly applicable model of group selection, the circumstances may still be limiting: "The extent to which this process of group selection occurs depends on (1) the validity of the trait-group concept, and given this, (2) the variation in the composition of trait-groups."

The development of the Price equation [29] allowed "…an ideal framework for addressing philosophical questions about levels of selection," [16] among other issues [30, 31]. Considerable discussion of inclusive fitness, reciprocity, group selection, and their role in the evolution of cooperation ensued (Fig. 3.1). Nevertheless, much of this discussion focused on modern, often multicellular, organisms. In the closing decades of the twentieth century, a few prescient minds began to recognize that, whatever the role of cooperation in modern life, its role in the history of life loomed larger [32, 33]. Indeed, much of the thinking about biological cooperation became strongly influenced by discussions of the history of life [16].

While reciprocity and kin selection can mediate evolutionary conflicts, in the history of life, numerous other mechanisms likely played a role in conflict mediation. These mechanisms typically decrease the variation of lower-level units, making the evolution of defectors less likely, or increase the variation of the higher-level units, thus allowing selection to favor cooperative groups, or both. As described by Michod [34] in the context of the evolution of multicellularity:

> …we define a conflict mediator as a feature of the cell group that restricts the opportunity for fitness variation at the lower level (cells) and/or enhances the variation in fitness at the higher level (the cell group or organism). Accordingly, one can think of two general classes of conflict mediators: those that restrict within-group change and those that increase the variation in fitness between groups, although both have the effect of increasingly the heritability of fitness at the group level.

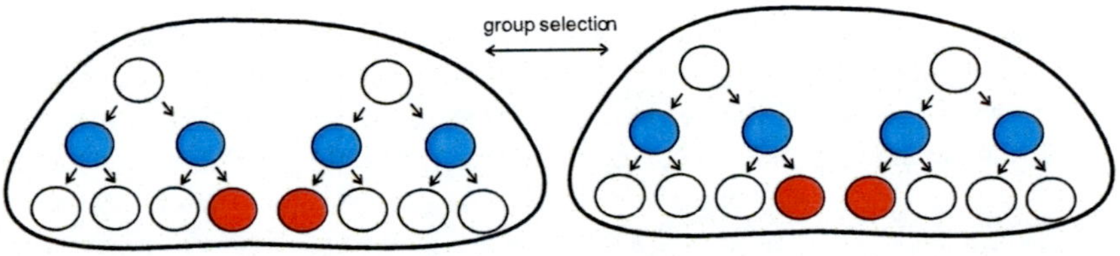

Fig. 3.1 Summary diagram showing kin selection (for instance, between shaded individuals in the second row) and reciprocity (for instance, between shaded individuals in the third row), while group selection occurs between the groups of individuals. In this example, reproduction is asexual, and arrows connect parent and offspring

By the end of the twentieth century, it became axiomatic that complex life requires cooperation. The capabilities of prokaryotes seem limited in this regard. As suggested by Conway Morris [35]: "The history of life shouts 'Look! Once there was bacteria, now there is New York'." While complexity and simplicity are always relative, collectively prokaryotes are metabolically versatile but structurally simple. Eukaryotes, on the other hand, while metabolically simple, are complex in terms of genes, genomes, structure, function, and life cycles. In many examples of eukaryotic cooperation, kin and group selection often work together. While the exact details remain obscure, eukaryotes are thought to have emerged from a syntrophic association among prokaryotes [36–38]. Metabolic reciprocity no doubt mediated conflict in these associations (Chap. 7). As described in Chap. 4, selection for cooperation has likely been a major factor in the history of life. In many ways, multicellular organisms can be regarded as a group of closely related cells that became eusocial, while eukaryotic cells can be regarded as a group of closely related (as well as unrelated) prokaryotes that became eusocial. We will return to these topics in subsequent chapters.

While not the focus of the current discussion, it is worth pointing out that recent studies of cooperation in modern life have become more focused on human behavior, which may be less affected by genes and the environment, and more affected by culture and tradition. Certainly, as suggested in Chap. 1, cooperation has a crucial role in human society. As pointed out by Chittka and Mesoudi [39]:

> Human groups are frequently not united by common interest in the way that honeybee swarms are united by shared kinship. The former often comprise conflicting factions each fighting for their own self-interest. And when human groups do act as cohesive units, they are often too cohesive, with their members rarely acting as independent decision-makers like honeybee scouts. Conformity prevents dissenting views and conflicting evidence from being considered....

Cultural factors (e.g., conformity) can thus lead to cooperation. It may or may not be an oversimplification to say that we are cooperative because when we are little children our parents teach us to conform. Favoring such conformity are thus the family and also religion [40]. Threats to conformity, on the other hand, might include the teaching of evolution, among others. Appropriately, and bringing us back to the start of this chapter, Wilson and Wilson [41] trace the foundations of studies of cooperation back to Darwin's writings. Finally, leaving the last words to Steven Pinker [5]:

> Social psychology, the science of how people behave toward one another, is often a mishmash of interesting phenomena that are "explained" by giving them fancy names. Missing is the rich deductive structure of other sciences, in which a few deep principles can generate a wealth of subtle predictions—the kind of theory that scientists praise as "beautiful" or "elegant." Trivers derived the first theory in social psychology that deserves to be called elegant. He showed that a deceptively simple principle—follow the genes—can explain the logic of each major kinds of human relationships....

Certainly, these studies of human beings and other species will continue to be an active area of research. Meanwhile, another deceptively simple principle—follow the electrons—will be applied to the puzzle of cooperation in subsequent chapters.

References

1. Margulis L, Sagan D (1986) Microcosmos. Summit, New York
2. Darwin C (1964) On the origin of species, 1st edn. Harvard University Press, Cambridge, MA
3. Darwin C (2004) Descent of man. Penguin Books, London
4. Herron JC, Freeman S (2014) Evolutionary analysis, 5th edn. Pearson, Boston, pp 1–850
5. Pinker S (2002) The blank slate. Penguin Books, New York
6. Bowles S (2009) Did warfare among ancestral hunter-gatherers affect the evolution of human social behaviors? Science 324:1293–1298
7. Dugatkin LA (2011) The prince of evolution. CreateSpace, Seattle
8. Mereschkowsky C (1905) Uber Natur und Ursprung der Chromatophoren im Pflanzenreiche. Biol Centralbl 25:593–604. (English translation in Martin W, Kowallik KV (1999) Eur J Phycol 34:287–295)
9. Mereschkowsky C (1910) Theorie der zwei Plasmaarten als Grundlage der Symbiogenesis, einer neuen Lehre von der Entstehung der Organismen. Biol Centralbl 30:278–288, 289–303, 322–374, 353–367. (English translation in Kowallik KV, Martin WF (2021) Biosystems 199:104281)
10. Sapp J, Carrapico F, Zolotonosov M (2002) Symbiogenesis: the hidden face of Constantin Merezhkowsky. Hist Philos Life Sci 24:413–440
11. Wallin IE (1927) Symbionticism and the origin of species. Williams & Wilkins, Baltimore, pp 1–171
12. Margulis L (1970) Origin of eukaryotic cells. Yale University Press, New Haven
13. Wright S (1931) Evolution in Mendelian populations. Genetics 16:97–159
14. Michod RE, Herron MD (2006) Cooperation and conflict during evolutionary transitions in individuality. J Evol Biol 19:1406–1409
15. Frank SA (2003) Repression of competition and the evolution of cooperation. Evolution 57:693–705
16. Okasha S (2006) Evolution and the levels of selection. Oxford University Press, Oxford, pp 1–263
17. Leigh EG Jr (2010) The group selection controversy. J Evol Biol 23:6–19
18. Wynne-Edwards VC (1962) Animal dispersion in relation to social behavior. Hafner, New York, pp 1–653
19. Maynard Smith J (1964) Group selection and kin selection. Nature 201:1145–1147
20. Williams GC (1966) Adaptation and natural selection. Princeton University Press, Princeton, pp 1–307
21. Hamilton WD (1964) The genetical evolution of social behavior. I. J Theor Biol 7:1–16
22. Haldane JBS (1932) The causes of evolution. Longman, Green, London
23. Trivers RL (2000) Obituary: William Donald Hamilton 1936–2000. Nature 404:828
24. Hamilton WD (1964) The genetical evolution of social behavior. II. J Theor Biol 7:17–52
25. Trivers RL (1971) The evolution of reciprocal altruism. Q Rev Biol 46:35–57
26. Trivers RL (2004) Mutual benefits at all levels of life. Science 304:964–965
27. Fehr E, Henrich J (2003) Is strong reciprocity a maladaptation? In: Hammerstein P (ed) Genetic and cultural evolution of cooperation. MIT Press, Cambridge, MA, pp 55–82
28. Wilson DS (1975) A theory of group selection. Proc Natl Acad Sci U S A 72:143–146
29. Price GR (1972) Extension of covariance selection mathematics. Ann Hum Genet 35:485–490. https://doi.org/10.1111/j.1469-1809.1957.tb01874.x
30. Lehtonen J, Okasha S, Helanterä H (2020) Fifty years of the Price equation. Philos Trans R Soc B 375:20190350. https://doi.org/10.1098/rstb.2019.0350
31. Shelton DE, Michod RE (2020) Group and individual selection during evolutionary transitions in individuality: meanings and partitions. Philos Trans R Soc B 375:20190364
32. Buss LW (1987) The evolution of individuality. Princeton University Press, Princeton
33. Szathmáry E, Demeter L (1987) Group selection of early replicators and the origin of life. J Theor Biol 128:463–486

34. Michod RE (2003) Cooperation and conflict mediation during the origin of multicellularity. In: Hammerstein P (ed) Genetic and cultural evolution of cooperation. MIT Press, Cambridge, MA, pp 291–307
35. Conway Morris S (2011) Complexity: the ultimate frontier? EMBO Rep 12:481–482
36. Martin W, Muller M (1998) The hydrogen hypothesis for the first eukaryote. Nature 392:37–41
37. Martin WF, Garg S, Zimorski V (2015) Endosymbiotic theories for eukaryote origin. Philos Trans R Soc B 370:20140330
38. Embley TM, Martin W (2006) Eukaryotic evolution, changes and challenges. Nature 440:623–630
39. Chittka L, Mesoudi A (2011) Insect swarm intelligence. Science 331:401–402
40. Wilson DS (2002) Darwin's cathedral. University of Chicago Press, Chicago
41. Wilson DS, Wilson EO (2007) Rethinking the theoretical foundation of sociobiology. Q Rev Biol 82:327–348

Chapter 4
A Thumbnail Sketch of the History of Life

The history of life is a history of different units of selection. Novel selective scenarios dominate at times of transition between units of selection. Whereas the lower self-replicating unit was previously selected by the external environment alone, following the transition it became selected by traits expressed by the higher unit. Variants expressed in the lower unit influence not only the relative replication rate of the lower unit, but also that of the higher unit. The potential clearly exists for variants to have a synergistic effect (that is, to favor the replication of both the lower and the higher unit), or for conflicts to arise. The organization of any unit will come to reflect those synergisms between selection at the higher and the lower levels which permit the new unit to exploit new environments and those mechanisms which act to limit subsequent conflicts between the two units. This explicitly hierarchical perspective on evolution predicts that the myriad complexities of ontogeny, cell biology, and molecular genetics are ultimately penetrable in the context of an interplay of synergisms and conflicts between different units of selection.

Leo Buss [1]

It is easy to see that the history of life exists conceptually just like human history. In both cases, certain events occurred at certain times in the past. In both cases, much of what occurs today remains deeply embedded in the matrix of earlier events. Despite clear parallels, there are also clear differences. The history of life occurred over a vastly larger tableau than human history. Human historical accounts usually begin more-or-less by convention about 6000 years ago [2]. Traces of life on Earth, on the other hand, extend nearly to the origin of the Earth itself [3–5]. For human history, an abundant archaeological record exists. Additionally, for at least some of human history, earlier historical accounts have been written down and preserved, although interpretation of these accounts may present challenges [6]. By contrast,

N. W. Blackstone, *Energy and Evolutionary Conflict*,
https://doi.org/10.1007/978-3-031-06059-5_4

evidence for the history of life is for the most part extremely sparse and fragmented by the very processes that produce it [7].

While it is easy to see that the history of life exists conceptually, it is thus much more difficult to actually reconstruct that history. Some recent discussions are rather pessimistic concerning the prospect of ever recovering such a history because of the process of horizontal gene transfer [8]. To see why horizontal gene transfer poses such a problem, we have to consider how the history of life is reconstructed using evolutionary trees.

Consider the case of the vast majority of modern evolutionary trees, in which nucleotide sequences constitute the shared derived characters that are used to unite groups [9]. If these sequences are vertically transmitted—i.e., passed from parent to offspring, like alleles for cooperation in Chap. 3—then their history will at least in part reflect the history of the cells and organisms that contain them. However, what if a nucleotide sequence from one organism colonizes and inserts into the genome of other cells and organisms? Such "horizontal" transmission is typically selected for at the level of the gene. Particularly in prokaryotes, ordinary genes frequently migrate between host cells [10]. The host cells themselves frequently benefit, acquiring new functional capabilities. These horizontally transmitted sequences will thus not accurately reflect the history of the cells that carry them because their own histories may be quite different. Further, it is not an easy task to distinguish vertical from horizontal inheritance simply by examining the nucleotide sequences. Phylogenetic trees based on nucleotide sequence data may thus be particularly unreliable for deep branches that occurred early in the history of life when prokaryotic cells promiscuously exchanged genes. Developing methods that compensate for horizontal transfer will better elucidate these ancient events [11].

Even when well supported, phylogenetic trees can only be informative up to the common ancestor of all life, variously termed the "most recent common ancestor" or the "last universal common ancestor" (LUCA). Between LUCA and the origin of life, there may lie a considerable gap that cannot be illuminated by the comparative method. Living things, however, likely contain some vestiges of life before LUCA. An intuitive understanding of vestiges can be gained by considering some aspects of human societies. For example, when Tony Blair was elected prime minister of the United Kingdom, he requested an audience with the Queen to ask permission to form a new government. On its face, such a request might seem a bit odd, since the Queen had no real authority to deny such permission. Only in the context of English and British history (e.g., parliament once served at the pleasure of the monarch) does this request begin to make sense. Similarly, modern cells may contain vestiges of their common history prior to LUCA. For instance, the replication of mobile genetic elements passes through an RNA stage. It may be that this stage and the many other things that RNA does in a cell (e.g., mRNA, tRNA, rRNA) are vestiges of an "RNA world" [12]. In the putative RNA world, RNA served as both template and catalyst, functions that are now more commonly carried out by DNA and proteins.

In addition to phylogenetic trees and vestiges, the geological record provides evidence of the history of life [13]. This record is based on the geological time scale

(Fig. 4.1). In an effort that foreshadowed the "big science" of the twentieth and twenty-first centuries, eighteenth- and nineteenth-century geologists used the guiding principles of "uniformitarianism" to produce the geological time scale. According to these principles, geological processes observed today on the modern Earth have operated throughout the Earth's history. Over very great time periods, the result of these processes is the geological record. For instance, observations of the bottom of a lake show that new sediment accumulates on top of old sediment. In time, the lake may fill with sediment and the sediment itself may become compacted into sedimentary rock. Erosion in turn may eventually expose this rock. The relative age of the layers of this exposed sedimentary rock would then reflect the timing of their accumulation—younger layers on top of older layers. This "principle of superposition" coupled with several other uniformitarian principles allowed geologists to reconstruct the relative time scale of the geological record.

In the twentieth century, scientists began to take advantage of the radioactivity of atoms to develop a chronological time scale for geological events. Use of a number of radioactive isotopes in this way allows spanning the entire geological time scale, as suggested by the chronological dates in Fig. 4.1. Radiometric age can be combined with relative age if, for instance, a volcanic ash layer or volcanic inclusions occur in an outcrop of sedimentary rock.

Once the geological time scale was elucidated, considerations of the history of life became possible. Probably the most striking pattern in the fossil record is the progression of life in terms of diversity and abundance [14]. Early on, there are only meager chemical traces of life [3–5]. Celebrated evidence [15] of complex cells in some of the oldest sedimentary rocks on Earth has been brought into serious question [16]. Abundant and convincing unicellular fossils are not found until much more recently (e.g., [13], but see [17, 18]). Most strikingly, unambiguous evidence of macroscopic, multicellular life only appears in the most recent one-sixth or so of the fossil record. The disappearance of these first "Ediacaran" creatures roughly corresponds to the beginning of the Phanerozoic—the eon of visible life. Only at this point in the Earth's history do macroscopic fossils become common and easily relatable at some level to extant living things. All of this is clear from examining the fossil record—not even the chronological time provided by radiometric dating is really necessary to see these patterns.

While these patterns have been clear since the formalization of the geological time scale, a conceptual understanding of these patterns has been curiously lacking. Once life originated, why would complex cells require such a long time to evolve, and why would the flowering of multicellularity require still longer? As articulated in the epigraph for this chapter, Buss [1] provides the conceptual framework to understand these questions: the history of life is a chronicle of evolutionary transitions in units or levels of selection. These ideas rapidly took root and reconceptualized the study of cooperation [19]. The study of the major transitions in the history of life was born [20–24]. As with all truly useful ideas, in hindsight this framework is intuitively obvious. Imagine being an observer near a rift vent at the bottom of the Archean Ocean. The first living things were likely groups of molecules. Perhaps after an initial stage in which energy was converted abiotically, these molecules

Fig. 4.1 The geological time scale was developed using several principles of uniformitarianism. Chronological time depends on radiometric dating. (Based on the Geological Time Scale, Wikipedia)

EON	ERA	Period	Epoch	AGE mya
Phanerozoic	Cenozoic	Quaternary	Holocene	
			Pleistocene	
		Neogene	Pliocene	
			Miocene	
		Paleogene	Oligocene	
			Eocene	
			Paleocene	66
	Mesozoic	Cretaceous		
		Jurassic		
		Triassic		252
	Paleozoic	Permian		
		Carbon-iferous		
		Devonian		
		Silurian		
		Ordovician		
		Cambrian		541
Proterozoic				2500
Archaean				4000
Hadean				4540

became capable of replication and energy conversion (Fig. 4.2). Copying errors led to variation then as now. Selection likely favored faster replication and more efficient energy conversion. Via variation and selection, some molecules made a discovery that was to resonate throughout the entire history of life. If a molecule relied on its sister molecules to carry out some or all of the tasks of energy conversion, this "selfish" molecule could then replicate at a higher rate and increase its frequency in the population (Fig. 4.2). Indeed, such a molecule could specialize on some aspects of living things—information and replication—at the expense of the others, energy conversion. The selfish molecule would be selected for as long as it did not become so abundant that the group's supply of energy became severely depleted. In this latter case, the entire group of molecules would be endangered and might go extinct. Some groups of molecules might evolve mechanisms that limited the extent to

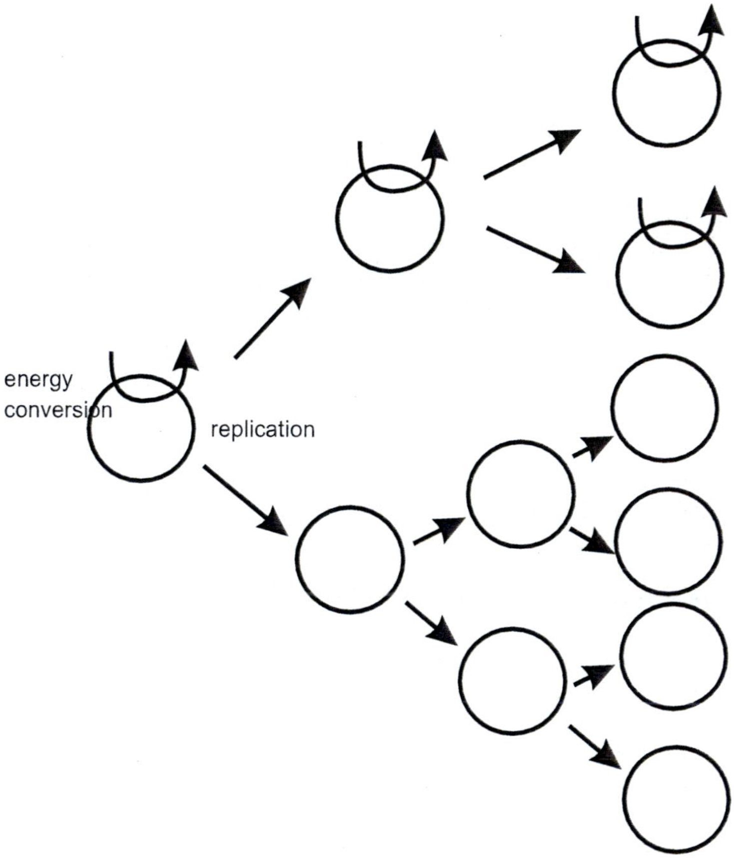

Fig. 4.2 Since life is characterized by replication and energy conversion, the typical pattern of conflict emerges in groups where resource sharing occurs. Replication specialists ("defectors") arise by loss-of-function mutations and forgo energy conversion in return for short-term gains in replication, but their success will be frequency dependent

which selfish molecules could plunder the group's energy stores. The groups that evolved such mechanisms would be more successful than groups that did not evolve them. For instance, the cell itself may be one such mechanism, protecting kin groups of replicators from exploitation by outside defectors. Horizontal gene transfer would in turn circumvent this protective function of the cell.

While the origin of life is one of the great mysteries of science, once life evolved this scenario of conflict between individual selfishness and the "good of the group" is inevitable. As soon as a common good exists, individuals will evolve to subvert it. The history of life is defined by these themes. Competition drives individuals to defect from the common good. In turn, successful groups evolve mechanisms to ensure cooperation of their constituent individuals. These groups emerge as higher-level units and go on to compete with other such emergent individuals and the cycle repeats itself. In this way, the complexity of the history of life evolved: genes within chromosomes, chromosomes within cells, cells within complex cells, and complex cells within multicellular organisms. Each major transition was a momentous event that required considerable evolution to accomplish. Simple cells, for instance, required transcendent evolution to produce complex cells. An incremental process of evolution could not yield this major transition. Symbiosis may be a powerful force in evolution [25], and symbioses between groups of prokaryotes may have occurred many times in the history of life [26], but only once did the necessary mechanisms of conflict resolution evolve and thus only once did complex, eukaryotic cells emerge. The lengthy gap between the first cells and the first eukaryotes in the fossil record thus likely reflects the stringent evolutionary requirements for such a step and the many failed attempts to fulfill them (see Chap. 7).

In modern biology, mobile genetic elements and viruses parallel these first replication specialists. Both are ubiquitous in all of cellular life [27, 28]. There are also numerous cellular organisms that at some level rely on a host organism to supply them with energy. Various terms can be used to describe the evolutionary dynamics of these sorts of host-symbiont interactions. "Cooperators" and "defectors" will be used here because these terms accurately describe the evolutionary process that occurs with multiple levels of selection. As pointed out by Nowak [29], the former forgo some of their reproductive potential to help the community, but natural selection inexorably favors the latter unless mechanisms of conflict mediation are at work. Other terms—e.g., parasites, cheaters, and selfish individuals—can be substituted more-or-less interchangeably for "defectors," while cooperators can be referred to as mutualists. The typical pattern of evolutionary conflict thus emerges, with selection at the lower level favoring defection and short-term gains in reproductive success, while selection at the higher level favors cooperation and long-term persistence.

Whether modern or ancient, the success of defectors is frequency dependent [23]. When they are rare, the products of energy conversion are abundant, and defectors replicate rapidly. As the proportion of defectors increases within a group, the availability of the energy currency decreases. At some point, the lack of products of energy conversion threatens the entire group. There would thus be strong selection at the level of the group to limit defectors (as well as strong selection on the defectors to migrate to a new group). Selection at the lower level favors shirking energy conversion tasks and specializing in replication and dispersal. Selection at the level of the group favors a balance between energy conversion and replication. Evolutionary conflict and its mediation are thus inherent to life. Mechanisms of conflict mediation that limit defection and favor cooperation are selected for at the level of the group. This pattern of conflict and conflict mediation has repeated itself again and again in the history of life. Lower-level units fail to convert or share energy. Higher-level units evolve mechanisms to ensure that they do both.

This much has been well said previously [1, 20–24, 30]. What has not been emphasized is the fundamental conflict inherent in the two essential aspects of life—replication and energy conversion. It seems inevitable that for any biological unit there will be functional trade-offs for these tasks. Put another way, the molecule or cell that specializes in replication will likely be somewhat impaired with regard to energy conversion and vice versa. Conversely, the molecule or cell that devotes itself equally to both tasks will be the proverbial jack of all trades and the master of none. Consequently, the lower-level unit that specializes in terms of replication can only plausibly gain an advantage by plundering the group's energy resources. Hence, it must be expected that the mechanisms that mediate levels-of-selection conflicts in favor of the higher-level unit must necessarily include those mechanisms that safeguard the group's energy resources against depredation by selfish individuals. In other words, metabolic regulation (see Chaps. 10 and 11) evolved at least in part to counter the effects of selfish replicators. A good part of the history of life may catalogue the evolution of just such mechanisms (Fig. 4.3). This can be viewed as a central tenet in the history of life.

Today, a multilevel view of selection is part of mainstream thinking in evolutionary biology. Any biological unit that exhibits heritable variation and is also selected will consequently evolve. As suggested above, living organisms may consist of several such units or levels. For instance, mobile genetic elements found in the genome of an organism may be transcribed into RNA. A protein enzyme called reverse transcriptase can then copy this RNA back into DNA, and the DNA can reinsert itself into the genome [27]. The original mobile genetic element has now increased its frequency relative to other elements of the genome. Selection at this level will favor this behavior. Hence, it is no wonder that these elements constitute a large part of animal genomes [28]. Meanwhile, the cell containing the mobile genetic element

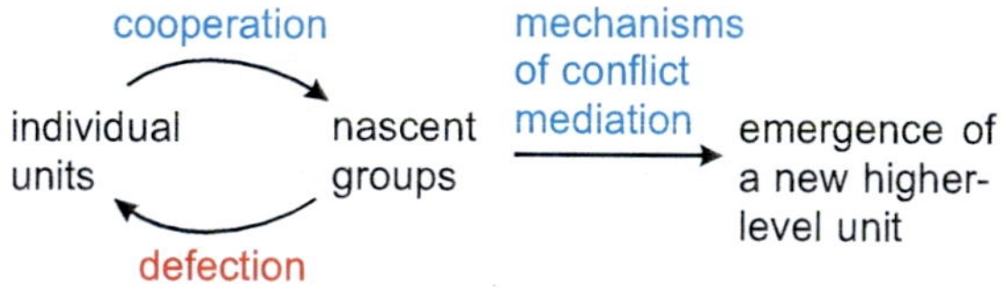

Fig. 4.3 Cycles of cooperation and conflict occur repeatedly in the history of life with individual biological units banding together to form groups and defecting individuals weakening the integrity of groups. For a higher-level biological unit to form, mechanisms of conflict mediation must be derived. (From Blackstone [31])

acquires a genetic mutation that allows it to continue to replicate instead of becoming terminally differentiated. This cell copies itself, forming a clone of daughter cells. Selection at the cell level favors this behavior. At the same time, the organism containing the cell and the mobile genetic element has encountered a rich supply of food. After gorging itself, the organism divides in half, and each half forms a complete organism that continues to feed and grow. Again, selection at the organism level will favor this behavior.

The preceding example not only illustrates multilevel selection but also illuminates a related concept—that of synergisms and antagonisms between levels of selection. In parallel to Darwin's example of human individuals within tribes (Chap. 3), the mobile genetic element could replicate itself in such a way that it might affect cell- or organism-level selection. Perhaps the mobile genetic element reinserted itself into the genome near a gene that is involved in the regulation of cell proliferation. Because of the action of the mobile genetic element, the cell containing it can now continue to replicate rather than becoming terminally differentiated. This exemplifies synergy between levels of selection. Both the cell and the mobile genetic element experience a gain in fitness. Meanwhile, the cell replicates to such an extent that organismal functions begin to suffer, e.g., the organism can no longer feed or replicate. This exemplifies antagonism between levels of selection. A trait selected for at both the level of the mobile genetic element and the level of the cell is nevertheless selected against at the level of the organism.

If the mobile genetic element and the cell are vertically transmitted (i.e., passed from parent to offspring organism), then selection on the higher-level unit will favor the evolution of any number of mechanisms to resolve this conflict in favor of the organism. Mobile genetic elements may be blocked from inserting into sensitive areas of the genome. The function of other genes may be co-opted into regulating cell proliferation. Policing by immune system cells may remove the hyperproliferative cells. Cells with mutations related to proliferation may commit suicide via programmed cell death. The organism itself may evolve to reproduce at an earlier age. All of this rich natural history can be interpreted and understood within a levels-of-selection framework, much as predicted by Buss [1] in the epigraph.

Such a scenario also has metabolic implications. As the selfish cell produces a clone of daughter cells, the growing tumor needs to gain access to the organism's food supply. Typically, this is done by triggering the growth of vascular tissue. Indeed, all human oncogenes and tumor-suppressor gene pathways have been implicated in angiogenesis, either directly or indirectly [32]. These cancer cells are extremely profligate with the organism's food resources. The well-known Warburg effect—the tendency for mammalian cancers to exhibit a glycolytic metabolism even in an aerobic environment [33–37]—can be understood in this context (but see [38, 39]). Glycolysis yields roughly one-tenth as much energy per unit of food as the much more efficient oxidative phosphorylation. Glycolysis does provide some advantages, however. First, compared to glycolysis plus the Krebs cycle (Chap. 2), it is fast and thus allows fast replication. Second, it allows the growing tumor to monopolize the organism's food supply, leaving less opportunity for other defecting cells to emerge and compete for resources in the somatic environment. Finally, glycolysis favors proliferation by producing precursors for biosynthesis.

Beyond these simple generalities, the history of life may be characterized by fundamental relationships between nutrient supply, metabolism, and cooperation that are only now beginning to come into focus. These topics will be further investigated in subsequent chapters.

References

1. Buss L (1987) The evolution of individuality. Princeton University Press, Princeton
2. Wilson DS (2007) Evolution for everyone. Delacorte Press, New York
3. Rosing MT (1999) ^{13}C-depleted carbon microparticles in >3700-Ma sea-floor sedimentary rocks from West Greenland. Science 283:674–676
4. Canfield DE (2006) Gas with an ancient history. Nature 440:426–427
5. Ueno Y, Yamada K, Yoshida N, Maruyama S, Isozaki Y (2006) Evidence from fluid inclusions for microbial methanogenesis in the early Archaean era. Nature 440:516–519
6. Flanagan T (1988) Tenants of time. Dutton, New York
7. Schindel DE (1980) Microstratigraphic sampling and the limits of paleontologic resolution. Paleobiology 6:408–426
8. Beiko RG, Doolittle WF, Charlebois RL (2008) The impact of reticulate evolution on genome phylogeny. Syst Biol 57:844–856
9. Goodman M (founding editor) (1992) Molecular phylogenetics and evolution. Elsevier, Amsterdam
10. Boc A, Philippe H, Makarenkov V (2010) Inferring and validating horizontal gene transfer events using bipartition dissimilarity. Syst Biol 59:195–211
11. Coleman GA, Davín AA, Mahendrarajah TA, Szánthó LL, Spang A, Hugenholtz P, Szöllősi GJ, Williams TA (2021) A rooted phylogeny resolves early bacterial evolution. Science 372:eabe0511. https://doi.org/10.1126/science.abe0511
12. Gilbert W (1986) Origin of life: the RNA world. Nature 319:618
13. Knoll AH (2003) Life on a young planet. Princeton University Press, Princeton
14. Bonner JT (1988) The evolution of complexity. Princeton University Press, Princeton
15. Schopf JW (1993) Microfossils of the early Archean apex chert: new evidence of the antiquity of life. Science 260:640–646
16. Brasier MD, Green OR, Jephcoat AP, Kleppe AK, Van Kranendonk MJ, Lindsay JF, Steele A, Grassineau NV (2002) Questioning the evidence for Earth's oldest fossils. Nature 416:76–81
17. Buick R (2010) Ancient acritarchs. Nature 463:885–886
18. Javaux EJ, Marshall CP, Bekker A (2010) Organic-walled microfossils in 3.2-billion-year-old shallow-marine siliciclastic deposits. Nature 463:934–938
19. Okasha S (2006) Evolution and the levels of selection. Oxford University Press, Oxford, pp 1–263
20. Szathmáry E, Demeter L (1987) Group selection of early replicators and the origin of life. J Theor Biol 128:463–486
21. Maynard Smith J, Szathmáry E (1995) The major transitions in evolution. Oxford University Press, Oxford, pp 1–346
22. Maynard Smith J, Szathmáry E (1999) The origins of life. Oxford University Press, Oxford
23. Michod RE (1999) Darwinian dynamics: evolutionary transitions in fitness and individuality. Princeton University Press, Princeton
24. Hammerstein P (ed) (2003) Genetic and cultural evolution of cooperation. The MIT Press, Cambridge, MA

25. Martin WF (2017) Physiology, anaerobes, and the origin of mitosing cells 50 years on. J Theor Biol 434:2–10
26. Martin WF, Garg S, Zimorski V (2015) Endosymbiotic theories for eukaryotic origin. Philos Trans R Soc Lond B 370:20140330
27. Hurst GDD, Werren JH (2001) The role of selfish genetic elements in eukaryotic evolution. Nat Rev Genet 2:597–606
28. Koonin EV, Makarova KS (2018) Anti-CRISPRs on the march. Science 362:156–157
29. Nowak MA (2006) Five rules for the evolution of cooperation. Science 314:1360–1363
30. Lane N (2015) The vital question: energy, evolution, and the origin of complex life. Norton, New York
31. Blackstone NW (2016) An evolutionary framework for understanding the origin of eukaryotes. Biology 5(2):18
32. Vogelstein B, Kinzler KW (2004) Cancer genes and the pathways they control. Nat Med 10:789–799
33. Garber K (2006) Energy deregulation: licensing tumors to grow. Science 312:1158–1159
34. Matoba S, Kang J-G, Patino WD, Wragg A, Boehm M, Gavrilova O, Hurley PJ, Bunz F, Hwang PM (2006) p53 regulates mitochondrial respiration. Science 312:1650–1653
35. Trachootham D, Zhou Y, Zhang H, Demizu Y, Chen Z, Pelicano H, Chiao PJ, Achanta G, Arlinghaus RB, Liu J, Huang P (2006) Selective killing of oncogenically transformed cells through a ROS-mediated mechanism by ß-phenylethyl isothiocyanate. Cancer Cell 10:241–252
36. Shaw RJ (2006) Glucose metabolism and cancer. Curr Opin Cell Biol 18:598–608
37. Kaelin WG Jr, Thompson CB (2010) Clues from cell metabolism. Nature 465:562–564
38. Chance B (2005) Was Warburg right? Or was it that simple? Cancer Biol Ther 4:125–126
39. Xu HN, Nioka S, Glickson JD, Chance B, Li LZ (2010) Quantitative mitochondrial redox imaging of breast cancer metastatic potential. J Biomed Opt 15(3):1–10

Chapter 5
Early Insights: A Fascination with Metabolic Gradients

Life depends on energy, in most cases on the energy that comes to the earth as solar radiation. Much of the activity of animals is devoted to getting the food which is their energy source. Discussions of the relative merits of different behavior patterns and life history strategies often pose the question, how can the animal best use its limited resources of energy? Energy is central to our understanding of animal life.

R. McNeill Alexander [1]

The evolutionary success of an organism depends on the extent to which it can extract energy from the environment, convert that energy into a useful form, and use at least some of it to make more of itself. Mechanisms of energy extraction and conversion would thus seem central to the study of evolution. Curiously, this has largely not been the case. While evolutionary biologists and ecologists accord immense importance to trophic competition between organisms [2] and energy allocation within an organism [3], the evolutionary biochemistry and physiology of mechanisms of energy acquisition and conversion have received considerably less attention (but see [4, 5]).

In part, this stems from a schism between the sub-disciplines of the biological sciences. Discussions of life and its origins typically focus on energy conversion, information, and replication [6–11]. Given this duality of life—energy metabolism versus information and replication—it might be expected that some disciplines would focus more on the former, while other disciplines would focus more on the latter. Indeed, this is the case with biochemistry and molecular biology. With roots in physiology, biochemistry focuses on studying chemical processes within living systems and particularly within cells. Molecular biology is a newer, hybrid discipline with roots perhaps best described by Francis Crick [12]:

> I myself was forced to call myself a molecular biologist because when inquiring clergymen asked me what I did, I got tired of explaining that I was a mixture of crystallographer,

N. W. Blackstone, *Energy and Evolutionary Conflict*,
https://doi.org/10.1007/978-3-031-06059-5_5

biophysicist, biochemist, and geneticist, an explanation which in any case they found too hard to grasp.

The real focus of molecular biology, however, is biological information [13]. Such information is of course the key to heredity. The search for the chemical nature of heredity in the 1940s and 1950s followed naturally from the "modern synthesis" of Darwin's theory of evolution and Mendelian genetics (see Chap. 3). Its discovery in one of the most celebrated scientific publications of all time [14] marked the advent of molecular biology. Evolutionary biology embraced this new discipline, since it was a natural outgrowth of the study of genes, information, and replication.

Other aspects of biology were ignored or perhaps even distained by the modern synthesis. In somewhat different contexts and with different emphases, a number of authors have previously recounted this history [15–20]. Here, I focus on a particular thread of this history that relates to energy metabolism. Indeed, by the late nineteenth century, many biologists attributed considerable effects to what were then rather vague notions of bioenergetics. One of these biologists was Charles Manning Child [20]. In 1900, Child began what was to become an extended series of experiments on regeneration using various animal taxa. In searching for processes governing regeneration, Child developed the idea of axial gradients. Libbie Hyman, widely known for her works on invertebrate zoology and during this time first a graduate student and then a research assistant in Child's laboratory [18, 19], provides a unique perspective [21]:

> About 1910 he [Child] began to perceive that the unity of the organism is a matter of correlation; his search for the mechanism of correlation led to the gradient theory which emerged about 1911 and with which his name will always be associated. The numerous researches he conducted on the regeneration of planarians from 1910 to 1915 led him to the concept of the existence in such simple axiate organisms of a gradation in rate of physiological processes along the axis. In this gradient Child believed he had found the mechanism of correlation by which the mass of cells that constitutes an animal is maintained as a unified whole of definite form and construction.

Beginning in about 1915, Child pursued a research program that demonstrated the existence of axial gradients in many different species. The essential methodology consisted of placing an organism in a solution of, for instance, potassium cyanide, known as an inhibitor of oxidative metabolism, and observing which parts of the organism were the first and which were the last to be inhibited. The gradient was thus revealed with metabolically active regions being the most susceptible and inactive regions being the least susceptible. Supporting data were provided by redox-sensitive stains, as well as studies of oxygen uptake of parts of cut-up organisms. Libbie Hyman performed most of the oxygen uptake experiments. At the time, criticisms of Child and co-workers focused on the use of metabolic toxins, on the methods used to study oxygen uptake (Winkler titrations), and on inferring oxygen uptake of the intake organism from cut-up parts. Nevertheless, as Child pointed out, the different aspects of the work were highly consistent, e.g., a gradient revealed by metabolic toxins was also found by redox-sensitive dyes and by measures of oxygen uptake.

Bonner [22] provides an insightful review of Child's conceptualization of metabolic gradients. In a developing biological system, one part becomes metabolically dominant by sequestering substrate needed for energy conversion. This part maintains dominance over other parts by continued success in competition for oxygen and food, and a gradient forms. Small quantitative differences in metabolic activity then result in qualitative differences in development. Nevertheless, such a gradient can only extend a finite distance, and beyond this distance, a new dominance-gradient system forms, based on the same principles.

Child's gradient theory was first widely applied to studies of regeneration. In the early twentieth century, a number of very able experimental biologists including Hans Driesch, Jacques Loeb, and Thomas Hunt Morgan were studying regeneration in addition to Child. As reviewed by Rose [23], the hydroid *Tubularia* became a favorite for these studies because of its large size and rapid rate of regeneration. *Tubularia* consists of a tubelike stalk or stem surrounded by perisarc and bearing a feeding hydranth at the distal end and a base at the proximal end (Fig. 5.1). A favorite experiment of the time was to remove the hydranth from the distal end and the base from the proximal end of a stem. The result was an undifferentiated tube surrounded by perisarc except for an opening at each end. Such a stem would quickly regenerate a hydranth at its distal end. However, regeneration at the distal end could be prevented by, for instance, burying that end in sand. Invariably, the other end of the stem would then regenerate the hydranth. Child's view was that the distal end had the highest metabolic rate. This end would continue to dominate the remainder of the stem and would regenerate the hydranth unless its metabolism was inhibited. The sand blocked the diffusion of oxygen; hence, the proximal end, adjacent to the opening in the perisarc, would then become dominant and regenerate the hydranth. A variety of other barriers to diffusion were employed with similar results.

Rose [23] relates the long series of experiments that eventually supported an alternative hypothesis—the sand was blocking the release of an inhibitory substance from the tissue of *Tubularia*. This morphogenetic substance then accumulated in the tissue and at sufficient concentrations inhibited the regeneration of the hydranth at that end. Thus, while subsequent researchers generally supported some of Child's hypotheses (e.g., the existence of metabolic gradients and dominance), the major difficulty of his larger theory was that of cause and effect. Clearly, much of development involves cellular processes (e.g., cell differentiation and movement) that necessarily entail energetic differences. Rapidly dividing or moving cells will no doubt consume more substrate and oxygen than quiescent cells. Do metabolic gradients cause developmental gradients or vice versa? Child could never adequately address this question [24].

Meanwhile, significant progress in biology was being made with more tractable problems. TH Morgan was one of Child's colleagues studying regeneration in the early twentieth century. In a dramatic change of direction that was to have profound consequences for biology particularly in the United States, after 1910 Morgan focused entirely on *Drosophila* and transmission genetics. Morgan's "Fly Room" subsequently became legendary and exerted a tremendous influence on the fledgling science of genetics. In part, this led to a divergence between studies of genetics and

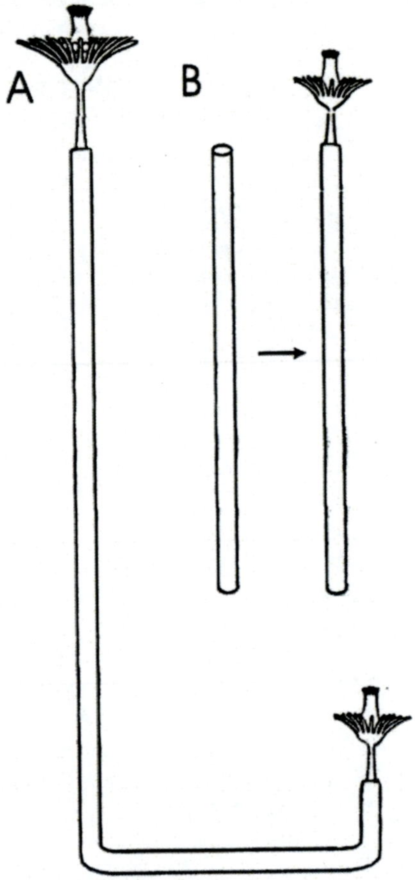

Fig. 5.1 (**a**) Stems of the hydroid *Tubularia* with two hydranths and a connecting stolon. The stolon encrusts the substratum and represents the proximal end or base, while the hydranth represents the distal end. (**b**) If both ends are severed, an undifferentiated tube results, which will typically regenerate a hydranth at its distal end. Various manipulations, however, can cause the proximal end to regenerate instead. Child suggested a metabolic basis for these patterns. (After Blackstone [20])

development [15, 17]. This divergence was in no small part caused by actual biological differences between *Drosophila* and organisms that Child worked on, such as *Tubularia* and planarian flatworms. Gilbert et al. [17] point out the relevance of these biological characteristics to the scientific debates between gradients and fields on one hand and transmission genetics on the other:

> In planaria, the inherited information could be seen in a gradient which enabled the organism to form a head at one end and a tail at the other. Upon splitting, each half inherited the ability to make a whole and properly organized animal. In *Drosophila*, several generations of flies could inherit a trait according to strict statistical laws, suggesting the involvement of nuclear chromosomes. The gene and the field were in opposition.

Much of the early success of genetics resulted from focusing on certain kinds of animals (those that determine their germ line in early embryology and only reproduce via gametes) while ignoring other animals [15]. Child and other advocates of the continued study of these "inconvenient" animals became increasingly ignored as well.

Larger social currents influenced this scientific dispute. Organisms capable of agametic, asexual reproduction tend to be highly responsive to environmental signals [25]. Generally, Child's worldview was very compatible with generalizations derived from such animals. Child accorded considerable weight to environmental influences in shaping human individuals [26]. Child argued that the individual was much more plastic than supposed by geneticists. This contrasted sharply with some interpretations of mechanistic genetics. Some criticism of the modern synthesis was based on a rejection of its social and political implications [27].

Despite the best efforts of Child and some other scientists, the triumph of mechanistic genetics was not to be denied. The discovery [14] of the chemical basis of heredity in 1953 and the advent of molecular biology emphasized and reinforced this success. Increasingly, the study of energy metabolism was ignored by evolutionary biologists in favor of the burgeoning field of molecular biology. The fields of biological information and replication became more and more accessible and available for exploitation by scientists in numerous studies. At the same time, while general ideas about metabolism and respiration had been part of scientific thinking for centuries, by the 1950s, no hypothesis could clearly explain the central process of energy conversion. There was no widely accepted mechanism linking respiration and the oxidation of carbon compounds to the synthesis of ATP, the cell's battery. Such a state of affairs was not likely to attract talented evolutionary biologists hoping to examine the role of energy conversion in the evolutionary process.

The central role of energy conversion in evolution actually facilitated the endeavors of evolutionary biologists who choose to ignore it. Strong selection on energy acquisition and conversion likely produced at least locally optimal solutions in the history of life that could not be readily improved on. In the groups of organisms studied by the architects of the modern synthesis, there was little variation in energy conversion efficiencies. As recounted by Buss [15]:

> Similarly, the natural historians central to the development of the Modern Synthesis were largely zoologists. Dobzhansky worked on *Drosophila*; Mayr, Simpson, and Rensch on birds and mammals. Dipterans and vertebrates were the areas of expertise of the individuals who framed the synthesis. This zoological bias is particularly revealing.

While Buss focuses on the zoological bias in mode of development, there is also another zoological bias in terms of energy metabolism. Insects and vertebrates are eukaryotes with mitochondria. Most of their cells' energy metabolism takes place in mitochondria. Further, insects and terrestrial vertebrates (particularly birds and mammals) are obligate aerobes, typically using diatomic oxygen as the terminal electron acceptor in oxidative phosphorylation. With relatively little mechanistic variation in the organisms of study [28, 29], energy conversion could be more-or-less ignored by the modern synthesis. Some of the best evolutionary studies of bioenergetics in insects reflect this by focusing not on oxidative phosphorylation in mitochondria, but on cytosolic, glycolytic enzymes [30]. Indeed, in terms of energy metabolism, eukaryotes themselves are remarkably homogeneous, as suggested by Müller et al. [29]:

> …anaerobic protists possess a few enzymes, about 10, that mammals do not. In contrast anaerobic metazoans survive by using the same enzymes humans do in their mitochondria albeit with parts of the Krebs cycle running backwards…

By and large, evolutionary studies could thus comfortably focus on replication and biological information, increasingly illuminated by the burgeoning field of molecular biology. As the modern synthesis became preeminent, however, the field of biochemistry underwent a renaissance of its own, revealing aspects of biology as fundamental as the chemical nature of heredity—namely, chemiosmosis, the chemical basis for energy conversion, as discussed in Chap. 2.

Child's focus on the significance of energy metabolism entirely predated the development of the chemiosmotic hypothesis. Had Child and colleagues been familiar with this theory, they may have made the sorts of connections that are outlined in subsequent chapters. With the discovery of chemiosmosis and quantum electron transfer, this early insight can be extended with a more complete view of energy and evolution. The relationship between multicellularity and metabolism will be revisited in a later chapter.

References

1. Alexander RM (1999) Energy for animal life. Oxford University Press, Oxford
2. Peiman KS, Robinson BW (2010) Ecology and evolution of resource-related heterospecific aggression. Q Rev Biol 85:133–158
3. Stearns SC (1992) The evolution of life histories. Oxford University Press, Oxford
4. Watt WB (1985) Bioenergetics and evolutionary genetics: opportunities for new synthesis. Am Nat 125:118–143
5. Watt WB (1986) Power and efficiency as indexes of fitness in metabolic organization. Am Nat 127:629–653
6. de Duve C (2002) Life evolving. Oxford University Press, Oxford
7. Martin W, Russell MJ (2003) On the origins of cells: a hypothesis for the evolutionary transitions from abiotic geochemistry to chemoautotrophic prokaryotes, and from prokaryotes to nucleated cells. Philos Trans R Soc Lond B 358:59–85
8. Maynard Smith J, Szathmáry E (1995) The major transitions in evolution. Oxford University Press, Oxford, pp 1–346
9. Maynard Smith J, Szathmáry E (1999) The orgins of life. Oxford University Press, Oxford
10. Morowitz HJ (1992) Beginnings of cellular life. Yale University Press, New Haven
11. Schrödinger E (1945) What is life? Cambridge University Press, Cambridge
12. Crick FHC (1965) Recent research in molecular biology: introduction. Br Med Bull 21:183–186
13. Stent GS (1968) That was the molecular biology that was. Science 160:390–395
14. Watson JD, Crick FHC (1953) Molecular structure of nucleic acids: a structure for deoxyribose nucleic acid. Nature 171:737–738
15. Buss L (1987) The evolution of individuality. Princeton University Press, Princeton
16. Allen RG, Balin AK (1989) Oxidative influence on development and differentiation: an overview of a free radical theory of development. Free Radic Biol Med 6:631–661

17. Gilbert SF, Opitz JM, Raff RA (1996) Resynthesizing evolutionary and developmental biology. Dev Biol 173:357–372
18. Maienschein J (1999) Libbie Hyman at the University of Chicago. Am Mus Novit 3277:25–32
19. Jenner RA (2004) Libbie Henrietta Hyman (1888–1969): from developmental mechanics to the evolution of animal body plans. J Exp Zool (MDE) 302B:413–423
20. Blackstone NW (2006) Charles Manning Child (1869–1954): the past, present, and future of metabolic signaling. J Exp Zool (MDE) 306B:1–7
21. Hyman LH (1955) Charles Manning Child: 1869–1954. Biogr Mem Natl Acad Sci 30:73–103
22. Bonner JT (1996) Sixty years of biology. Princeton University Press, Princeton
23. Rose SM (1970) Regeneration. Appleton-Century-Crofts, New York
24. Blackstone NW (1998) Individuality in early eukaryotes and the consequences for metazoan development. Prog Mol Subcell Biol 19:23–43
25. Blackstone NW, Bridge DM (2005) Model systems for environmental signaling. Integr Comp Biol 45:605–614
26. Mitman G, Fausto-Sterling A (1992) Whatever happened to Planaria? C. M. Child and the physiology of inheritance. In: Clarke AE, Fujimura JH (eds) The right tools for the job. Princeton University Press, Princeton, pp 172–197
27. Davis EB (2005) Science and religious fundamentalism in the 1920s. Am Sci 93:253–260
28. Bryant C (ed) (1991) Metazoan life without oxygen. Chapman and Hall, New York
29. Müller M, Mentel M, van Hellemond JJ, Henze K, Woehle C, Gould SB, Yu R-Y, van der Giezen M, Tielens AGM, Martin WF (2012) Biochemistry and evolution of anaerobic energy metabolism in eukaryotes. Microbiol Mol Biol Rev 76:444–495
30. Wheat CW, Watt WB, Pollock DD, Schulte PM (2006) From DNA to fitness differences: sequences and structures of adaptive variants of *Colias* phosphoglucose isomerase (PGI). Mol Biol Evol 23:499–512

Chapter 6
How Can Metabolism Lead to Groups?

One of the most studied and most poorly understood concerns of ethology is why animals live in groups. Living in groups has been claimed to aid in the rearing of young, to facilitate mating, to increase foraging success, to reduce the risk of predation, to provide protection from inclement weather, and to increase swimming efficiency.

Daniel Rubenstein [1]

At the risk of a tiresome repetition of the obvious, group-living organisms, whether of the same or different species, have more opportunities for cooperation than those that live solitary lives. Indeed, theories for the evolution of cooperation typically assume some sorts of social interactions as a matter of course. Wilson [2] defined sociobiology as "the systematic study of the biological basis of all social behavior. For the present it focuses on animal societies, their population structure, castes, and communication, together with all the physiology underlying the social adaptations." Sociobiology takes a group-structured population more-or-less as a given. As suggested by the epigraph, there are many hypotheses as to why animals live in groups. Of particular interest here, some of these hypotheses involve the distribution of resources and the transmission dynamics of symbionts [3, 4].

Meanwhile, by the late twentieth century, microbiologists increasingly adopted the "biofilm" model of natural microbial communities. In contrast to laboratory cultures in which planktonic cells may be selected, biofilms include multispecies assemblages attached to a surface and surrounded by a thick layer of extracellular polymeric substances [5]. According to Flemming and Wingender [6], these substances, "immobilize biofilm cells and keep them in close proximity, thus allowing for intense interactions, including cell–cell communication, and the formation of synergistic microconsortia."

No doubt, group living facilitates social interactions in communities from microbes to mammals. Group formation itself should thus not be ignored when considering the evolution of cooperation. This should also be kept in mind when

N. W. Blackstone, *Energy and Evolutionary Conflict*,
https://doi.org/10.1007/978-3-031-06059-5_6

considering major transitions in evolution, where it might be tempting to assume a group-structured population at the outset. Such an assumption may result in skipping ahead to the middle of an evolutionary transition, rather than starting at the beginning. Important steps may inadvertently be missed by such an approach. As pointed out by Radzvilavicius and Blackstone [7]: "To the extent that individuality is an emergent trait, the problem can be approached by recognising the importance of individuating mechanisms that are present from the very beginning of the transition, when only lower-level selection is acting."

The impact of metabolism in general and chemiosmosis in particular may be most strongly felt in these nascent steps of a transition. There are several, related pathways by which metabolism can lead to groups. Consider syntrophy, or "feeding together," which dominates many microbial communities [8]. Community formation may be based on syntrophic relationships, in which the waste of one microbe can become the substrate of another (Fig. 6.1). Symbiosis based on syntrophy can be viewed as a form of reciprocity, as suggested by Trivers [9]. In some cases, evolutionary conflict can be further mediated by simple stoichiometry. For instance, if one microbe produces a reduced compound as waste (XH in Fig. 6.1), its energy conversion reactions will be enhanced by any process that removes that waste. The lowered concentration from such a removal process will alleviate end-product inhibition [10], which otherwise would slow or otherwise disrupt energy conversion. Thus, another microbe that takes up XH and oxidizes it to X would indeed be a valuable partner. Further, the first microbe could reciprocate by taking up the oxidized compound, X, lowering its concentration, and alleviating end-product inhibition in its partner. Defection from this partnership has automatic negative consequences in the form of a buildup of waste and a shortage of substrate. Stoichiometry thus provides a form of automatic conflict mediation in syntrophic associations of this sort.

Importantly, the same processes that result in conflict mediation also drive group formation. In the above example, one can imagine the "unaffiliated" partners, each releasing what could be substrate for the other while suffering some degree of end-product inhibition and perhaps shortages of substrate. The partners would naturally be attracted to one another, ensuring a source of substrate as well as a sink for unwanted product. The nature of the metabolic processes themselves also remains

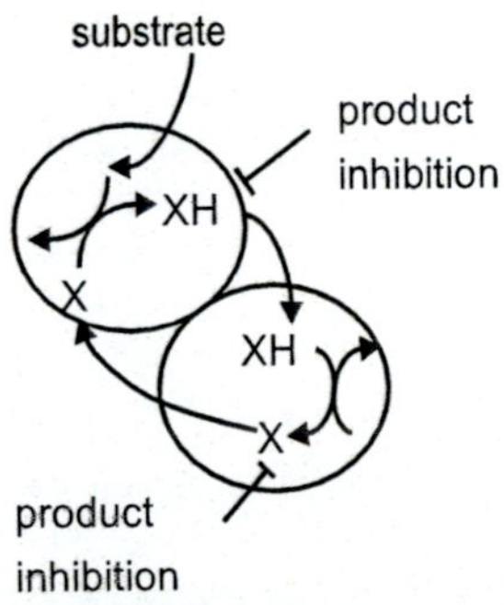

Fig. 6.1 Syntrophy can result in communities of microbes. Further, conflict can be alleviated by reciprocity. Each partner alleviates end-product inhibition by taking up the other's waste product and producing more of its reactants

unspecified. Both partners could be chemiosmotic or both fermentative, or there could be one of each.

In this context, it is worthwhile to examine the "hydrogen hypothesis," a well-known exemplar of metabolic associations that may have figured prominently in the origin of eukaryotes [11]. In an anaerobic environment, heterotrophic, sometimes aerobic, bacteria carry out glycolysis but need to re-oxidize NADH and do this by converting pyruvate to CO_2, H_2O, H_2 (the electrons from NADH are funneled onto H^+ to make H_2), and fatty acids. These waste products can be used by autotrophic archaea where H_2 is now the electron donor and CO_2 is reduced both to sugars and to methane, which is released as waste. In time, the association may have become intimate, with the bacterial community eventually becoming endosymbiotic. It may seem that the bacteria are initially useful to the archaea but not vice versa, i.e., that this is a commensalism. When end-product inhibition is taken into consideration, however, the initial advantages to the bacteria become clear, and properly this should be regarded as a mutualistic symbiosis.

Nevertheless, introducing chemiosmosis into syntrophic relationships adds a layer of complexity that is likely absent from purely fermentative interactions. First, rates of reaction should be considered. While glycolysis is considered "fast" relative to complete processing of substrate in the Krebs cycle and chemiosmosis, note that the latter includes glycolysis, i.e., in the metabolism of glucose, glycolysis feeds substrate into the Krebs cycle. As discussed in Chap. 2, because of electron tunneling and supercomplex formation, purely chemiosmotic reactions proceed at higher rates than purely fermentative ones. These reaction-rate imbalances likely require differential investment in enzymatic machinery in mixed partnerships. Further, there is a large difference in the costs of end-product inhibition in chemiosmotic and fermentative reactions. For instance, human cells can carry out glycolysis when deprived of oxygen for brief periods. The accumulation of lactic acid causes pH to drop, eventually inhibiting the glycolytic enzymes. On the other hand, as already discussed in Chap. 2, inhibiting mitochondria with insufficient metabolic demand and an excess of substrate has considerably more dire consequences: the production of reactive oxygen species (ROS) [12]. The same effect occurs with chloroplasts that are exposed to light, but which lack electron acceptors [10]. While chemiosmosis can be modulated to some extent [13, 14], cells that rely on chemiosmosis might be expected to be more prone to syntrophic associations that remove excess product in exchange for a supply of reactants.

Consider a cyanobacterial cell that carries out oxygenic photosynthesis in the marine environment. With abundant light and water, it can produce quantities of ATP and NADPH. These can be used to fix carbon dioxide. Since a cell does not have infinite storage capacity, at some point, there will be more reduced carbon, in whatever form, than the cell can store. The cell can then take steps to modulate photosynthesis, or simply export the excess product to the external environment, or both. There will be strong selection to take some action, since doing nothing will result in end-product inhibition, highly reduced electron carriers, and increased ROS formation. Exporting reduced carbon can attract microbes that utilize this substrate in energy conversion. This may regenerate carbon dioxide, which may or may

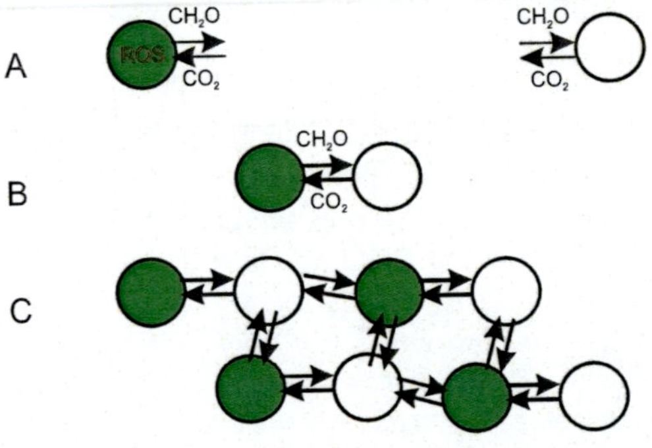

Fig. 6.2 A cyanobacterium (shaded) under ideal circumstances for photosynthesis exports reduced carbon to alleviate end-product inhibition and reactive oxygen species (ROS) formation (**a**). Another microbe imports reduced carbon and exports carbon dioxide. A mutually beneficial symbiosis develops (**b**). Individuals multiply, forming a cooperative community (**c**)

not be in short supply, and it can lower the concentration of reduced carbon, simplifying its export. In many ways, a typical syntrophic association may develop (Fig. 6.2).

Nevertheless, this association is asymmetric: foremost, the cyanobacterium simply needs to get rid of reduced carbon. With high levels of bicarbonate typically present in seawater, carbon dioxide is unlikely to be in short supply, and reduced carbon may be dispersed by water movement. The cyanobacterium reaps a benefit simply by dispersing inconvenient amounts of product into the environment. As a by-product of this necessary step [15], a symbiosis may form. Again, the first step in this symbiosis would be the formation of groups, which may be temporary (e.g., during daylight hours) or more permanent. As discussed in Chap. 3, groups facilitate cooperation either by reciprocity, kin selection, group selection, or all three.

As with other syntrophic interactions, the characteristics that drive group formation can also mediate conflict. Consider a symbiosis such as that described above. Once a community of cooperators has formed, a defecting cyanobacterium might monopolize substrate rather than exporting it. Under some circumstances, this may allow the defector to replicate faster. Under the circumstance that led to the symbiosis, however, monopolizing substrate may lead to end-product inhibition, highly reduced electron carriers, and ROS. Chemiosmosis not only leads to group formation but enforces cooperation once groups form.

In summary, the hypothesis that chemiosmosis can mediate conflict and lead to associations among organisms is based on three premises [16]:

1. Under circumstances that favor chemiosmosis, the metabolic needs of a cell or organism may be easily met because this process is extraordinarily fast and efficient.
2. By its biochemical nature—separating hydrogen atoms into component electrons and protons—chemiosmosis can be a potentially fraught process.

3. Given (1), in chemiosmotic cells and organisms, too much ATP (or other product) is more frequently a problem than too little. Given (2), too much ATP (or other product) can be risky in that it leads to end-product inhibition causing loose electrons to form dangerous by-products.

Cooperation is usually selected against because of the evolutionary costs of sharing, but if chemiosmosis diminishes these costs, or removes them entirely, or even converts costs into benefits, cooperation can then be favored and associations among organisms can form. Nevertheless, even in groups of cooperators formed in this manner, defectors may still arise, e.g., by mutations that counteract passive "leakage" of product through a cell wall, or by loss-of-function mutations to genes coding for transporters that would otherwise carry excess product out of the cell or organism. If the excess product can then be diverted into greater reproductive success, defection will be favored. If, however, defection leads to end-product inhibition, it will have costs and will be selected against. Thus, if the origin of life is "the free lunch you are paid to eat" [17], then when the lunch is no longer free, you must make your lunch or steal someone else's. With chemiosmosis, however, there may be "the free lunch you are forced to make" [16].

These aspects of metabolism in general and chemiosmosis in particular have largely been ignored in considerations of symbiosis, whether in major transitions in the history of life or in modern taxa. While group formation has sometimes been taken for granted in considerations of the evolution of cooperation, it may indeed be a key step. The metabolic drivers of group formation may also have a role in mediating evolutionary conflicts that arise subsequent to group formation. Considerations of metabolism provide a more complete picture of the costs and benefits of symbiosis and multicellularity. As mentioned in Chap. 1, on sunny days, plants provide mycorrhizal fungi with abundant photosynthate. Is this a cost or a benefit? Similarly, on sunny days, corals with symbiotic algae release large amounts of nutrient-rich material into the ocean. Cost or benefit? When defecting from the good of the multicellular group, mammalian cancer cells famously downregulate chemiosmosis when exhibiting the so-called Warburg effect. Again, cost or benefit? In the next several chapters, these ideas will be applied to major transitions (the origin of eukaryotes and multicellularity) as well as to these and other modern symbioses.

References

1. Rubenstein DI (1978) On predation, competition, and the advantages of group living. In: Bateson PG et al (eds) Social behavior. Plenum Press, New York
2. Wilson EO (1975) Sociobiology. Harvard University Press, Cambridge, MA
3. Johnson DPD, Kays R, Blackwell PG, Macdonald DW (2002) Does the resource dispersion hypothesis explain group living? Trends Ecol Evol 17:563–570
4. Lombardo MP (2008) Access to mutualistic endosymbiotic microbes: an underappreciated benefit of group living. Behav Ecol Sociobiol 62:479–497
5. Costerton JW (1999) Introduction to biofilm. Int J Antimicrob Agents 11:217–221
6. Flemming H-C, Wingender J (2010) The biofilm matrix. Nat Rev Microbiol 8:623–633

7. Radzvilavicius AL, Blackstone NW (2018) The evolution of individuality, revisited. Biol Rev 93:1620–1633
8. Morris BEL, Henneberger R, Huber H, Moissl-Eichinger C (2013) Microbial syntrophy: interaction for the common good. FEMS Microbiol Rev 37:384–406
9. Trivers RL (1971) The evolution of reciprocal altruism. Q Rev Biol 46:35–57
10. Goldschmidt EE, Huber SC (1992) Regulation of photosynthesis by end-product accumulation in leaves of plants storing starch, sucrose, and hexose sugars. Plant Physiol 99:1443–1448
11. Martin W, Muller M (1998) The hydrogen hypothesis for the first eukaryote. Nature 392:37–41
12. Chance B, Williams GR (1956) The respiratory chain and oxidative phosphorylation. Adv Enzymol Relat Subj Biochem 17:65–134
13. Allen JF, Santabarbara S, Allen CA, Puthiyaveetil S (2011) Discrete redox signaling pathways regulate photosynthetic light-harvesting and chloroplast gene transcription. PLoS One 6:e26372
14. Malone LA, Qian P, Mayneord GE, Hitchcock A, Farmer DA, Thompson RF, Swainsbury DJK, Ranson NA, Hunter NA, Johnson MP (2019) Cryo-EM structure of the spinach cytochrome b6f complex at 3.6 Å resolution. Nature 575:535–539
15. Bronstein JL (ed) (2015) Mutualism. Oxford University Press, Oxford
16. Blackstone NW (2020) Chemiosmosis, evolutionary conflict, and eukaryotic symbiosis. In: Kloc M (ed) Symbiosis: cellular, molecular, medical, and evolutionary aspects. Springer, Cham, pp 237–252
17. Lane N (2009) Life ascending: the ten great inventions of evolution. Oxford University Press, Oxford

Chapter 7
Chemiosmosis and the Origin of Eukaryotes

> *Life arose around half a billion years after the earth's formation, but then got stuck at the bacterial level of complexity for more than 2 billion years, half the age of our planet. Indeed, bacteria have remained simple in their morphology (but not their biochemistry) throughout 4 billion years. In stark contrast, all morphologically complex organisms—all plants, animals, fungi, seaweeds and single-celled 'protists' such as amoeba—descend from that singular ancestor about 1.5–2 billion years ago.*
>
> *Nick Lane [1]*

Nick Lane's monumental book, *The Vital Question: Energy, Evolution, and the Origin of Complex Life*, delves into what may have been the most challenging of all the major evolutionary transitions: the origin of eukaryotes. Certainly, all evidence suggests a long period in which life on Earth was dominated by prokaryotes. What were the obstacles that hindered this transition for 2 billion years? While we may never completely comprehend the answers to this question, there can be no doubt that evolutionary conflict was a serious hurdle. Metabolism in general and chemiosmosis in particular likely had crucial roles in mediating these conflicts.

The advantages of eukaryotic cells are straightforward. Not only are they more complex, but they are also larger than their prokaryotic forebearers. As Bonner [2] points out, "…the reason for non-stop selection for organisms of increased size is that the top of the size scale is an ever-present open niche and has been open during the entire course of organic evolution." Larger size provides a number of ecological advantages including the exploitation of more and different food resources, more efficient dispersal (e.g., escaping the constraints of low Reynolds numbers), producing more offspring, and escaping predators [3]. Prokaryotes, however, face a conundrum that greatly limits their options for size increase: their energy-converting complexes are found on the cell membrane [4]. If a prokaryotic cell gets larger, there is less surface to convert energy and more volume requiring energy conversion

N. W. Blackstone, *Energy and Evolutionary Conflict*,
https://doi.org/10.1007/978-3-031-06059-5_7

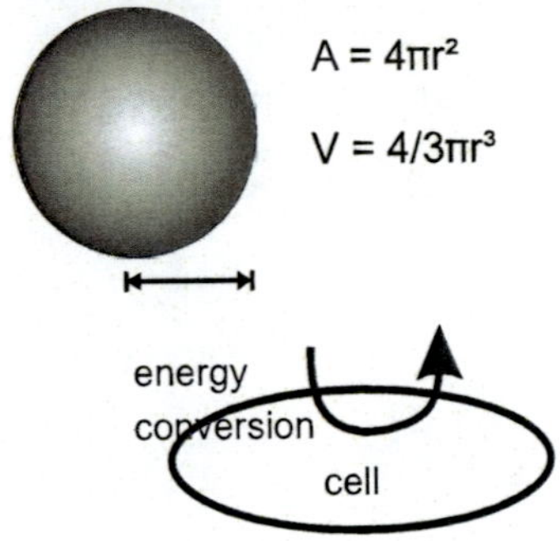

Fig. 7.1 Modeling a cell as a sphere, its surface area (A) increases as the square of the radius, while its volume (V) increases as the cube. Thus, as a prokaryotic cell gets larger, it has less surface for energy conversion and more volume requiring energy. (From Torday et al. [5])

(Fig. 7.1). Despite some exotic exceptions, by-and-large prokaryotes never transcended these surface-to-volume constraints.

One way to circumvent S-V constraints is to move small energy-converting cells inside a larger complex cell, thus freeing the external membrane from duties related to energy conversion [4, 6]. This arrangement allows the complex cell to increase in size. From this perspective, endosymbiosis is integral to the evolution of eukaryotes [7]. A clever engineering solution for surface-to-volume constraints, however, results in a levels-of-selection nightmare.

As described in previous chapters, all life takes up energy from the environment and typically converts this energy into more useful forms. At the same time, life involves information and replication. There remains a constant tension between these two attributes in all forms of life. In a structured population, a biological unit that relies on its sister units for energy conversion and specializes in replication may be favored. The success of such a unit, however, is frequency dependent [8]. As the proportion of units specializing in replication increases within a group, the availability of the products of energy conversion inexorably decreases. At some point, the lack of these products threatens the entire community. For the community to persist, mechanisms of conflict mediation must evolve.

While these issues prevail throughout the entire history of life, they were particularly acute in the origin of the eukaryotic cell. Two great symbioses—that of the mitochondrion and the plastid—figure prominently in the rise of eukaryotes. Both of these symbioses involve energy-converting lower-level units, with groups of these lower-level units and host cells constituting the higher-level units. In both cases, evolutionary conflict no doubt involved the usual pattern of lower-level units specializing in replication at the expense of distributing the products of energy conversion to the higher-level community.

As pointed out in Chap. 3, early-twentieth-century formulations of the endosymbiont theory of the origin of eukaryotes explicitly rejected "Darwinian" notions of conflict, posing cooperation as an alternative. Much the same thinking pervaded Lynn Margulis' writings as she resurrected the endosymbiont theory. While Margulis' ideas received considerable criticism [9–11], remarkably, none of this criticism was based on considerations of the evolution of cooperation. This is particularly surprising, because as described in Chap. 2, in the 1960s and 1970s, there was near-absolute condemnation of any kind of group-selection thinking in evolutionary theory [12]. Ultimately, abundant nucleotide sequence data convincingly showed that mitochondria and plastids were endosymbiotic relatives of alpha-proteobacteria and cyanobacteria, respectively [7]. Since some eukaryotes primitively lack plastids, their origin remains somewhat less mysterious than that of mitochondria [13, 14]. Indeed, the origin of mitochondria has been and continues to be one of the most debated topics in all of biology. Considerations have turned to the nature of the host and metabolic relationships that may have led to the endosymbiosis, topics that continue to attract considerable attention [15–20].

Nevertheless, there remained very little discussion of the evolutionary conflicts inherent in the origin of eukaryotes until the closing years of the twentieth century. Symbiosis, intimate relationships between different organisms, often at the cellular level, can range from parasitism to commensalism to mutualism. The last, in which both the host and the symbiont benefit, has attracted considerable attention. Why should individuals of two different species cooperate? Indeed, as pointed out above, such examples were thought to be contrary to Darwinian evolution [21]. Currently, however, explicit considerations of evolutionary conflict are recognized as a central issue in understanding symbioses [22], and cooperation is not an automatic outcome. In other words, as discussed in more detail below, even when a host-symbiont community appears to be dominated by mutualistic interactions, evolutionary conflict can still arise. Cooperation emerges if individuals forgo reproduction to contribute to the group, but selection will inevitably favor the opposite [23]. Mechanisms must evolve, or more typically be co-opted into mediating these evolutionary conflicts. These mechanisms can involve reciprocity, kin selection, or any number of idiosyncratic mechanisms that affect variation at different levels of selection. The last typically decrease variation at the lower level, making the evolution of defectors less likely, or increase variation at the higher level, allowing selection to favor the more cooperative groups, or both [24].

Early attempts to apply this framework to the mitochondrial endosymbiosis were largely ignored [25, 26]. Rather, the recognition of conflictual stages in the early evolution of eukaryotes grew out of empirical findings that showed a role for mitochondria in programmed cell death [27–30]. In general, the mitochondrial endosymbiosis now fits comfortably within the multilevel theory of evolution. Mitochondrial ancestors and hosts banded together into nascent groups. In many of these groups, conflict overpowered cooperation and the lower-level units returned to the free-living state. Yet in one lineage, the group derived mechanisms of conflict mediation, and a new higher-level unit—the eukaryotic cell—emerged. These mechanisms of conflict mediation likely constitute many of the shared derived features of eukaryotes [31–35].

Mitochondria may have been attendant to the very early stages in eukaryotic origins and were very likely a feature of the last eukaryotic common ancestor (LECA) [14]. Plastids were acquired by some eukaryotes not long after LECA [13]. In modern eukaryotes, evolutionary conflict is mediated by a number of mechanisms. Nevertheless, the evolution of many of these mechanisms required strong selection on the higher-level units. Such selection cannot have molded the initial steps in the symbiosis, which by definition involved only lower-level units. Notably, mitochondria, chloroplasts, and their bacterial relatives convert energy using chemiosmosis, which involves electron flow and proton extrusion.

As described in Chap. 2, chemiosmosis describes the ubiquitous process that most living organisms use to convert energy. Notably, it yields considerably more ATPs than energy conversion based on substrate-level phosphorylation (e.g., fermentation). Chemiosmosis also proceeds by mechanisms wholly different from substrate-level phosphorylation. First described in a series of revolutionary papers by Peter Mitchell [36], the surviving theory has been substantially modified from that which was originally proposed [37]. Nevertheless, the basic elements of chemiosmosis remain clearly recognizable [38]. Electron carriers, embedded in a membrane that is impervious to protons, connect what are ultimately environmental sources and sinks of electrons. As redox reactions proceed, the electron carriers translocate protons across the membrane. These protons then move back across the membrane via ATP synthase triggering the formation of ATP from ADP and inorganic phosphate. While microbial chemiosmotic processes are many and various, here the focus will be on oxidative phosphorylation in mitochondria and oxygenic photosynthesis in chloroplasts. While of course chloroplasts require light, in many ways, they function similarly to mitochondria. Electrons (from water in the former or coenzymes such as NADH in the latter) power an electron transport chain, producing a proton gradient, which catalyzes the formation of ATP in both chloroplasts and mitochondria (NADPH is also formed in the former) (Fig. 7.2). Cells containing mitochondria can then store ATP as phosphoenolpyruvate or phosphocreatine or something similar, while chloroplasts store the energy in ATP and NADPH by

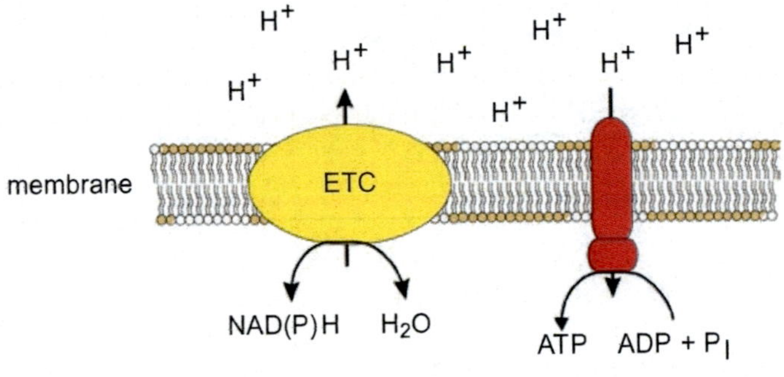

Fig. 7.2 Schematic summarizing eukaryotic chemiosmosis. Mitochondria oxidize reduced cofactors such as NADH, run the electrons through an electron transport chain (ETC), and build a transmembrane proton gradient. Protons return via ATP synthase, triggering the formation of ATP from ADP and inorganic phosphate (Pi). With the input of light energy, chloroplasts oxidize water, run the electrons through an electron transport chain (which is homologous to that in mitochondria), and build a transmembrane proton gradient. As in mitochondria, protons trigger the formation of ATP and electrons reduce NADP+ to NADPH. Mitochondria can store ATP as phosphoenolpyruvate, phosphocreatine, or similar compounds, while chloroplasts store the energy in ATP and NADPH by fixing carbon via the soluble enzyme RuBisCO. (From Blackstone [39])

fixing carbon via the soluble enzyme RuBisCO (ribulose-1,5-bisphosphate carboxylase/oxygenase).

At the same time that Mitchell was developing the chemiosmotic theory, other work began to reveal the existence of quantum electron transfer in biological systems [40]. In addition, later work showed the existence of "supercomplexes" among the membrane-bound electron carriers. Thus, electron transfer within and between membrane-bound complexes in chemiosmosis occurs extremely rapidly [41, 42]. This rapidity poses problems in linking chemiosmosis to the soluble reactions that store energy. For instance, RuBisCO is perhaps the most abundant protein on Earth because it is "mopping up" the products of chemiosmosis. The linking of chemiosmosis to slower soluble reactions and potentially limited storage capacity has other consequences as well. If an accumulation of products inhibits electron flow, these electrons may divert to molecular oxygen and reactive oxygen species (i.e., partially reduced forms of oxygen, ROS) will form. The chemiosmotic process itself is the cause of ROS formation.

In evolutionary terms, cooperation usually involves costs and thus is not an automatic outcome. Hosts and symbionts may respond to divergent selective forces [22]. A defecting symbiont can be selected to sequester resources from the host and symbiont community and to use these resources for its replication. Such defectors may gain a replicatory advantage compared to cooperative symbionts that at least in part forgo reproduction and share resources with the larger community. On the other hand, by sharing resources with the host, the cooperative symbiont community often establishes a more durable environment for their long-term persistence. Despite these long-term advantages, the host and the larger symbiont community (the higher-level unit) remain vulnerable to exploitation by lower-level defectors. While symbiosis is often conceptualized in bilateral terms (e.g., mutualism or parasitism), mechanistically these evolutionary interactions are multilateral and multilevel. In other words, even when a host-symbiont community appears to be dominated by mutualistic interactions, defecting symbionts can still arise and flourish unless they are controlled by mechanisms of conflict mediation. Population structure often plays a role in conflict mediation, particularly if a population is subdivided into many small groups. Such groups can potentiate kin selection and reciprocity, and groups of cooperators can arise purely by chance. In this way, even if cooperation is selected against at the level of the individual, it can still arise and be favored at the level of the group [43].

Remarkably, the process of chemiosmosis favors conflict mediation. Chemiosmosis proceeds rapidly and conserves a large proportion of the energetic input, quickly generating products. These products can be stored in various ways, but storage mechanisms are slow relative to chemiosmosis and in any event storage capacity is usually limited. When conditions are opportune, chemiosmotic cells and organisms face the possibility of "end-product inhibition" [44, 45], which can have severe consequences. An overabundance of product can inhibit electron flow. In the presence of molecular oxygen, this enhances the formation of reactive oxygen species. Such partially reduced forms of oxygen can have a variety of detrimental effects. To avoid blocking electron flow, the abundant products of chemiosmotic energy conversion must be consumed, stored, or simply gotten rid of. While mechanisms that modulate chemiosmosis are available [46, 47], an alternative solution is simply to disperse excess product into the environment.

Crucially, chemiosmosis and its consequences can thus under some circumstances favor sharing rather than hoarding the products of energy conversion. Such inadvertent largesse can lead to the formation of symbiotic associations, which are in some ways similar to "by-product" symbioses [22]. The resulting groups may be the key to the evolution of cooperation. Indeed, the biophysical constraints of chemiosmosis may have provided the initial mediation of conflicts associated with origins of eukaryotic cells and thus allowed the first steps in the formation of complex life on Earth.

As endosymbioses became established and higher-level units emerged, energy-converting lower-level units remained an obstacle to cooperation. Numerous additional mechanisms of conflict mediation subsequently evolved, facilitated by the large populations of higher-level units, which each contained relatively small populations of lower-level units. Purely on the basis of chance, some higher-level units could be formed from cooperative lower-level units. These groups of cooperators could then outcompete cells that included one or many defectors. Nevertheless, under conditions that lead to end-product inhibition, defectors that arise via loss-of-function mutations could still be eliminated from cooperative groups. In this way, unicellular eukaryotes developed robust and stable symbioses.

The biophysics of chemiosmosis, however, were likely central to the initial steps of eukaryogenesis, when only lower-level selection was operating. Quantum electron transfer and supercomplex formation drive chemiosmotic processes at extremely high rates, rapidly producing large quantities of products. While chemiosmosis can be modulated by several mechanisms, releasing excess product provides an alternative means to protect against end-product inhibition and a buildup of dangerous by-products such as reactive oxygen species. This "no-cost" sharing—the free lunch you are forced to make—facilitates interspecific groups, and such groups can lead to cooperative symbioses. Even after such groups have formed, however, cooperators are always vulnerable to exploitation by defectors, so additional mechanisms of conflict mediation are usually necessary. Under some, but not all, circumstances, chemiosmosis may further mediate the conflict.

Interestingly, the first rigorous reconstruction of the ancestral character states of eukaryotes suggests a strong possibility that LECA had a multinucleate stage in its life cycle [14]. Such a stage would enhance evolutionary conflict because larger groups weaken between-group and strengthen within-group selection. Chemiosmosis-related mechanisms of conflict mediation might have a greater role in such multinucleate forms, as might as-yet-unexplored mechanisms to mediate conflict between nuclei. On the other hand, given that the analysis also strongly suggests that LECA was sexual, unicellular gametes may have provided considerably variation for selection to have favored cooperators over defectors.

With this background, the success of chemiosmotic bacteria in the origin of eukaryotes can be better understood. The origin of the mitochondrion remains shrouded in mystery, but the evidence suggests that it occurred concomitantly with the origin of eukaryotes [1, 6, 16]. Initially, free-living protomitochondria may have passively released high-energy phosphate compounds such as pyrophosphate, perhaps in this way paralleling modern photosynthetic symbionts such as those involved in lichen symbioses (see Chap. 8). By doing so, end-product inhibition was avoided, and redox homeostasis maintained. Further, such largesse may have led to the formation of groups of microbes, perhaps including those that became the host in the partnership that led to eukaryotes. The initial symbiosis may have been based on an exchange of high-energy and low-energy phosphate compounds (Fig. 7.3). As the eukaryotic cell developed, ADP/ATP carriers evolved, perhaps initially acting as uncouplers to dissipate excess membrane potential [48]. Similarly, when exchanging ATP for ADP, these carriers simulate metabolic demand and alleviate end-product inhibition [43]. Again, while the origin of mitochondria and eukaryotes is an ancient event in the history of life, it is plausible that chemiosmosis had a twofold impact: triggering the formation of groups that led to a structured population and mediating conflict in favor of cooperators once defectors arose in these groups.

The events surrounding the origin chloroplasts are somewhat clearer, particularly in view of the recent reconstruction of ancestral character states [13]. Much like modern lichens or dinoflagellate-animal symbioses, the original symbiosis likely involved the exchange of reduced and oxidized carbon (Fig. 7.3). Chemiosmosis may have driven the association between early eukaryotes and cyanobacteria. As the

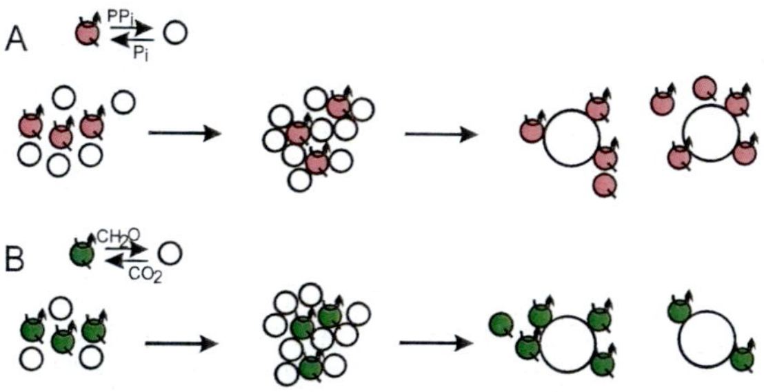

Fig. 7.3 Schematic outlining of the possible role of chemiosmosis in the symbioses that gave rise to the eukaryotic cell. In (**a**), protomitochondria (shaded circles) carry out oxidative phosphorylation, indicated by curved arrow. At times, these cells emit excess product, indicated by the diagonal line, perhaps in the form of pyrophosphate (PP_i). The proto-host cells (unfilled circles) take up pyrophosphate and emit inorganic phosphate (P_i). The evolutionary dynamics of the protomitochondria could have included cells that cease to carry out oxidative phosphorylation and take up pyrophosphate as well as cells that continue to carry out oxidative phosphorylation and cease to emit pyrophosphate. At times, the latter could be disadvantaged by end-product inhibition and reactive oxygen formation. In (**b**), proto-chloroplasts (shaded circles) carry out oxygenic photosynthesis, indicated by the curved arrow. At times, these cells emit excess product, indicated by the diagonal line, in the form of reduced carbon (CH_2O). The eukaryotic proto-host cells (unfilled circles) take up reduced carbon and emit oxidized carbon (CO_2). The evolutionary dynamics of the proto-chloroplasts could include cells that cease to carry out oxygenic photosynthesis and take up reduced carbon as well as cells that continue to carry out oxygenic photosynthesis and cease to emit reduced carbon. At times, the latter could be disadvantaged by end-product inhibition and reactive oxygen formation. (From Blackstone [39])

symbiosis evolved, mutations that inactivated the export of reduced carbon from chloroplasts may have led to defectors. Coupled with limitations on replication, end-product inhibition, and ROS formation may have punished defectors and enforced cooperation. The secondary symbioses between eukaryotes that primitively possessed chloroplasts and those that lacked them may have proceeded similarly.

Chemiosmosis may thus have had a powerful impact on the two most consequential symbioses in the history of life. It may also impact many modern symbioses particularly those that parallel the origin of mitochondria and chloroplasts: chemiosmotic symbionts releasing products to maintain redox homeostasis, thus leading to the formation of groups, and as the symbiosis develops, chemiosmosis being co-opted into further mediating evolutionary conflict (Chap. 8). Chemiosmosis may thus be one of the key drivers of the origin of eukaryotes, and its effects should likely be considered when drawing evolutionary parallels between modern symbioses and events that occurred earlier in the history of life.

References

1. Lane N (2015) The vital question: energy, evolution, and the origin of complex life. Norton, New York
2. Bonner JT (1998) The origins of multicellularity. Integr Biol 1:27–36
3. Lachmann M, Blackstone NW, Haig D, Kowald A, Michod RE, Szathmáry E, Werren JH, Wolpert L (2003) Group 3: Cooperation and conflict in the evolution of genomes, cells, and multicellular organisms. In: Hammerstein P (ed) Genetic and cultural evolution of cooperation. MIT Press, Cambridge, MA, pp 327–356
4. Lane N (2005) Power, sex, suicide: mitochondria and the meaning of life. Oxford University Press, Oxford
5. Torday JS, Blackstone NW, Rehan VK (2019) Evidence-based evolutionary medicine. Wiley, Hoboken
6. Lane N, Martin W (2010) The energetics of genome complexity. Nature 467:929–934
7. Martin WF (2017) Physiology, anaerobes, and the origin of mitosing cells 50 years on. J Theor Biol 434:2–10
8. Michod RE (1999) Darwinian dynamics. Princeton University Press, Princeton
9. Raff RA, Mahler HR (1972) The non-symbiotic origin of mitochondria. Science 177:575–582
10. Uzzell T, Spolsky C (1974) Mitochondria and plastids as endosymbionts: a revival of special creation? Am Sci 62:334–343
11. Bogorad L (1975) Evolution of organelles and eukaryotic genomes. Science 188:891–898
12. Williams G (ed) (1971) Group selection. Aldine Atherton, Chicago
13. Sánchez-Baracaldo P, Raven JA, Pisani D, Knoll AD (2017) Early photosynthetic eukaryotes inhabited low-salinity habitats. Proc Natl Acad Sci U S A 114:E7737–E7745
14. Skejo J, Garg SG, Gould SB, Hendriksen M, Tria FDK, Bremer N, Franjević D, Blackstone NW, Martin WF (2021) Evidence for a syncytial origin of eukaryotes from ancestral state reconstruction. Genome Biol Evol 13 (in press)
15. Embley TM, Martin W (2006) Eukaryotic evolution, changes and challenges. Nature 440:623–630
16. Martin WF, Garg S, Zimorski V (2015) Endosymbiotic theories for eukaryote origin. Philos Trans R Soc B 370:20140330

17. Spang A, Saw JH, Jørgensen SL, Zaremba-Niedzwiedzka K, Martijn J, Lind AE, van Eijk R, Schleper C, Guy L, Ettema TJG (2015) Complex archaea that bridge the gap between prokaryotes and eukaryotes. Nature 521:173–179
18. Speijer D (2017) Alternating terminal electron-acceptors at the basis of symbiogenesis: how oxygen ignited eukaryotic evolution. BioEssays 39:1600174
19. Dance A (2021) The mysterious microbes at the root of complex life. Nature 593:328–330
20. Liu Y, Makarova KS, Huang W-C, Wolf YI, Nikolskaya AN, Zhang X, Cai M, Zhang C-J, Xu W, Luo Z, Cheng L, Koonin EV, Li M (2021) Expanded diversity of Asgard archaea and their relationships with eukaryotes. Nature 593:553–557
21. Blackstone NW (2016) An evolutionary framework for understanding the origin of eukaryotes. Biology 5:18
22. Bronstein JL (ed) (2015) Mutualism. Oxford University Press, Oxford
23. Nowak MA (2006) Five rules for the evolution of cooperation. Science 314:1560–1563
24. Michod RE (2003) Cooperation and conflict mediation during the origin of multicellularity. In: Hammerstein P (ed) Genetic and cultural evolution of cooperation. MIT Press, Cambridge, MA, pp 291–307
25. Cosmides LM, Tooby J (1981) Cytoplasmic inheritance and intragenomic conflict. J Theor Biol 89:83–129
26. Blackstone NW (1995) A units-of-evolution perspective on the endosymbiont theory of the origin of the mitochondrion. Evolution 49:785–796
27. Frade JM, Michaelidis TM (1997) Origin of eukaryotic programmed cell death—a consequence of aerobic metabolism. BioEssays 19:827–832
28. Kroemer G (1997) Mitochondrial implication in apoptosis: towards an endosymbiont hypothesis of apoptosis evolution. Cell Death Differ 4:443–456
29. Mignotte B, Vayssiere J-L (1998) Mitochondria and apoptosis. Eur J Biochem 252:1–15
30. Blackstone NW, Green DR (1999) The evolution of a mechanism of cell suicide. BioEssays 21:84–88
31. Blackstone NW (2013) Why did eukaryotes evolve only once? Genetic and energetic aspects of conflict and conflict mediation. Philos Trans R Soc Lond B 368:20120266
32. Blackstone NW (2013) Evolution and cell physiology. 2. The evolution of cell signaling from mitochondria to Metazoa. Am J Physiol Cell Physiol 305:C909–C915
33. Blackstone NW (2014) sAC as a model for understanding the impact of endosymbiosis on cell signaling. Biochim Biophys Acta 1842:2548–2554
34. Blackstone NW (2015) The impact of mitochondrial endosymbiosis on the evolution of calcium signaling. Cell Calcium 57:133–139
35. Radzvilavicius AL, Blackstone NW (2015) Conflict and cooperation in eukaryogenesis: implications for the timing of endosymbiosis and the evolution of sex. J R Soc Lond Interface 12:20150584
36. Mitchell P (1961) Coupling of phosphorylation to electron and hydrogen transfer by a chemiosmotic type of mechanism. Nature 191:144–148
37. Boyer PD, Chance B, Ernster L, Mitchell P, Racker E, Slater EC (1977) Oxidative phosphorylation and photophosphorylation. Annu Rev Biochem 46:955–1026
38. Allen JF (1993) Control of gene expression by redox potential and the requirement for chloroplast and mitochondrial genomes. J Theor Biol 165:609–631
39. Blackstone NW (2020) Chemiosmosis, evolutionary conflict, and eukaryotic symbiosis. In: Kloc M (ed) Symbiosis: cellular, molecular, medical, and evolutionary aspects. Springer, Cham, pp 237–252
40. Chance B, Nishimura M (1960) On the mechanism of chlorophyll-cytochrome interaction: the temperature insensitivity of light-induced cytochrome oxidation in chromatium. Proc Natl Acad Sci U S A 46:19–24
41. Dudkina NV, Eubel H, Keegstra W, Boekema EJ, Braun H-P (2005) Structure of a mitochondrial supercomplex formed by respiratory-chain complexes I and III. Proc Natl Acad Sci U S A 102:3225–3229

42. Moser CC, Keske JM, Warncke K, Farid RS, Dutton PL (1992) Nature of biological electron transfer. Nature 355:796–802
43. Radzvilavicius AL, Blackstone NW (2018) The evolution of individuality, revisited. Biol Rev 93:1620–1633
44. Chance B, Williams GR (1956) The respiratory chain and oxidative phosphorylation. Adv Enzymol Relat Subj Biochem 17:65–134
45. Goldschmidt EE, Huber SC (1992) Regulation of photosynthesis by end-product accumulation in leaves of plants storing starch, sucrose, and hexose sugars. Plant Physiol 99:1443–1448
46. Allen JF, Santabarbara S, Allen CA, Puthiyaveetil S (2011) Discrete redox signaling pathways regulate photosynthetic light-harvesting and chloroplast gene transcription. PLoS One 6:e26372
47. Malone LA, Qian P, Mayneord GE, Hitchcock A, Farmer DA, Thompson RF, Swainsbury DJK, Ranson NA, Hunter NA, Johnson MP (2019) Cryo-EM structure of the spinach cytochrome b6f complex at 3.6 Å resolution. Nature 575:535–539
48. Bertholet AM, Chouchani ET, Kazak L, Angelin A, Fedorenko A, Long JZ, Vidoni S, Garrity R, Cho J, Terada N, Wallace DC, Spiegelman BM, Kiricho Y (2019) H+ transport is an integral function of the mitochondrial ADP/ATP carrier. Nature 571:515–520

Chapter 8
Chemiosmosis and Modern Symbioses

Mutualisms, interactions between two species that benefits both of them, have long captured the public imagination. Humans are undeniably attracted by the idea of cooperation in nature. For thousands of years, we have been seeking explanations of its occurrence in other organisms, often imposing our own motivations and mores in an effort to explain what we see. Spectacular natural history stories abound; some of them are even true.

Judy Bronstein [1]

Symbiosis, intimate relationships between different organisms, often at the cellular level, can range from parasitism to commensalism to mutualism. The last, in which both the host and the symbiont benefit, has attracted considerable attention. Why should individuals of two different species cooperate? Indeed, such examples seem contrary to Darwinian evolution and for some time were supposed to be just that [2]. Currently, however, explicit considerations of evolutionary conflict are recognized as a central issue in understanding symbioses [1]. In other words, as discussed in more detail below, even when a host-symbiont community appears to be dominated by mutualistic interactions, evolutionary conflict can still arise. Cooperation emerges if individuals forgo reproduction to contribute to the group, but selection will inevitably favor the opposite [3]. Mechanisms must evolve, or more typically be co-opted into mediating these evolutionary conflicts [4].

In evolutionary terms, cooperation usually involves costs and thus is not an automatic outcome. Hosts and symbionts may respond to divergent selective forces [1]. A defecting symbiont can be selected to sequester resources from the host and symbiont community and to use these resources for its replication. Such defectors may gain a replicatory advantage compared to cooperative symbionts that at least in part forgo reproduction and share resources with the larger community. On the other hand, by sharing resources with the host, the cooperative symbiont community often establishes a more durable environment for their long-term persistence.

N. W. Blackstone, *Energy and Evolutionary Conflict*,
https://doi.org/10.1007/978-3-031-06059-5_8

Despite these long-term advantages, the host and the larger symbiont community (the higher-level unit) remain vulnerable to exploitation by lower-level defectors. Cooperation can emerge only if mechanisms of conflict mediation evolve to suppress defectors. Kinship, reciprocity, and a number of idiosyncratic traits may do this. These mechanisms may decrease the heritable variation at the lower level, thus limiting the evolution of potential defectors, or increase the heritable variation at the higher level, thus potentiating selection against groups of defectors [4].

While symbiosis is often conceptualized in bilateral terms (e.g., mutualism or parasitism), mechanistically these evolutionary interactions are multilateral and multilevel. In other words, even when a host-symbiont community appears to be dominated by mutualistic interactions, defecting symbionts can still arise and flourish unless they are controlled by mechanisms of conflict mediation. Population structure often plays a role in conflict mediation, particularly if a population is subdivided into many small groups. Such groups can favor kin selection and reciprocity, and groups of cooperators can arise purely by chance. In this way, even if cooperation is selected against at the level of the individual, it can still arise and be favored at the level of the group [5].

There has been little consideration of energy metabolism in this context. As reviewed in Chap. 2, chemiosmotic reactions produce a transmembrane "proton-motive force" that powers energy-requiring reactions in most organisms. Further, under some circumstances, chemiosmosis may function as a mechanism of conflict mediation. As outlined in previous chapters, chemiosmotic reactions are extremely fast and can quickly produce large quantities of products. These products can be stored in various ways, but storage mechanisms are slow relative to chemiosmosis and in any event storage capacity is usually limited. When environmental conditions are favorable, chemiosmotic cells and organisms face the possibility of "end-product inhibition" [6, 7], which can have severe consequences. In some sense, chemiosmosis confronts organisms with the same issues that southern California electric utilities face on sunny, windy days—the need to entice consumers to use more electricity before transmission lines melt [8]. While mechanisms that modulate chemiosmosis are available [9, 10], an alternative solution is simply to disperse excess product into the environment. Such inadvertent largesse can lead to the formation of symbiotic associations, which are in some ways similar to "by-product" symbioses [1]. The resulting groups may be the key to the evolution of cooperation.

While microbial chemiosmotic processes are many and various, here the focus will for the most part be on eukaryotes, and thus oxidative phosphorylation in mitochondria and oxygenic photosynthesis in chloroplasts. While of course chloroplasts require light, in many ways, they function similarly to mitochondria. Electrons (from water in the former or coenzymes such as NADH in the latter) power an electron transport chain, producing a proton gradient, which catalyzes the formation of ATP in both chloroplasts and mitochondria (NADPH is also formed in the former). Cells containing mitochondria can then store ATP as phosphoenolpyruvate or phosphocreatine or something similar, while chloroplasts store the energy in ATP and NADPH by fixing carbon via the soluble enzyme RuBisCO (ribulose-1,5-bisphosphate carboxylase/oxygenase).

Chemiosmosis differs markedly from chemical reactions mediated by soluble enzymes. Remarkably, complexes of the electron transport chain employ quantum electron transfer [11]. Further, "supercomplexes" among the membrane-bound electron carriers can form. Thus, electron transfer within and between membrane-bound complexes in chemiosmosis occurs extremely rapidly [12, 13]. This rapidity poses problems in linking chemiosmosis to the soluble reactions that store energy. For instance, RuBisCO is perhaps the most abundant protein on Earth because it is "mopping up" the products of chemiosmosis. The linking of chemiosmosis to slower soluble reactions and potentially limited storage capacity has other consequences as well. If an accumulation of products inhibits electron flow, these electrons may divert to molecular oxygen and reactive oxygen species (i.e., partially reduced forms of oxygen, ROS) will form. The chemiosmotic process itself is the cause of ROS formation.

As recently summarized by Blackstone [14], the hypothesis that chemiosmosis can mediate conflict and lead to associations among organisms is based on three premises:

1. Under circumstances that favor chemiosmosis, the energetic needs of a cell or organism may be easily met because this process is extraordinarily fast and efficient.
2. By its biochemical nature—separating hydrogen atoms into component electrons and protons—chemiosmosis can be a potentially fraught process.
3. Given (1), in chemiosmotic cells and organisms too much ATP (or other product) is more frequently a problem than too little. Given (2), too much ATP (or other product) can be risky in that it leads to end-product inhibition causing loose electrons to form dangerous by-products.

Cooperation is usually selected against because of the evolutionary costs of sharing, but if chemiosmosis diminishes these costs, or removes them entirely, or even converts them into benefits, cooperation can then be favored and associations among organisms can form. Nevertheless, even in groups of cooperators formed in this manner, defectors may still arise, e.g., by mutations that counteract passive "leakage" of product through a cell wall, or by loss-of-function mutations to genes coding for transporters that would otherwise carry excess product out of the cell or organism. If the excess product can then be diverted into greater reproductive success, defection will be favored. If, however, defection leads to end-product inhibition, it will have costs and will be selected against. Thus, if the origin of life is "the free lunch you are paid to eat" [15], then when the lunch is no longer free, you must make your lunch or steal someone else's. With chemiosmosis, however, there may be "the free lunch you are forced to make."

Casual observation suggests that many successful symbioses involve an exchange of products produced by chemiosmosis. As described in Chap. 7, the symbioses that produced the eukaryotic cell involve chemiosmotic organelles—mitochondria and chloroplasts. Since all eukaryotes primitively contain mitochondria, eukaryotic symbioses typically involve photosynthesis. Secondary symbioses between non-photosynthetic and photosynthetic eukaryotes have occurred several times. Fungi and algae bond together in lichens; many marine animals rely on photosynthetic dinoflagellates; photosynthetic plants are symbiotic with nitrogen-fixing bacteria and arbuscular mycorrhizal fungi; insects feed on sap from photosynthetic plants; and so on. There is also a large class of associations between gut microbiota and metazoans—e.g., ruminants, termites, and even human beings—which at least in part include fermentative microbes releasing substrate that is utilized by the chemiosmotic host. In all of these symbioses, chemiosmosis figures prominently in the interaction. Is this merely a coincidence? Although existing symbioses cannot provide direct evidence into their formative steps, specific examples can still be instructive in terms of which sort of relationships—e.g., chemiosmotic symbionts or chemiosmotic hosts—favor mutualistic symbioses. By no means is the following intended to be a comprehensive review; rather, these examples are introduced to highlight themes that are then synthesized. The first example, that of corals and dinoflagellates, will be reviewed in detail, followed by brief mention of several others.

Coral-Dinoflagellate Symbioses While numerous marine animals form symbiotic associations, clonal and colonial animals such as sponges, ascidians, bryozoans, and cnidarians are particularly likely to do so [16–19]. Notably, all modern reef-building cnidarians contain endosymbiotic dinoflagellates [20], formerly considered *Symbiodinium* and now classified as the family Symbiodiniaceae [21]. Many other colonial cnidarians, whether part of coral reef communities or not, also exhibit similar symbioses. The coral-dinoflagellate symbiosis has attracted considerable study because its breakdown triggers coral bleaching. When environmental stress becomes extreme, these dinoflagellates are lost, and corals bleach [22]. As elaborated below, chemiosmosis likely alleviates evolutionary conflicts, but also contributes to the process of coral bleaching in which cooperation breaks down.

Notably, while taxa included in the Symbiodiniaceae form symbioses with corals and many other metazoans, they also remain capable of free-living existence. Given the intense competition for space in the marine benthos, symbiosis is a path by which Symbiodiniaceae can become larger and thus more effective competitors. For the metazoan host, symbiosis is a path to at least partial autotrophy, since these dinoflagellates are photosynthetic and actively export various forms of reduced carbon [22].

Despite these mutual benefits, a durable symbiosis requires robust mechanisms of conflict mediation. As suggested above, a population structure of many, small groups can often mediate a conflict. No matter how strongly defectors are selected for at the individual level, with many, small groups, purely by chance (i.e., genetic drift) some groups will comprise only cooperators. These groups of cooperators will then be strongly selected for at the group level and outcompete groups with more defectors. This sort of scenario likely contributed to the secondary symbioses that gave rise to dinoflagellates, among others [5]. Some stages of the life cycle of colonial cnidarians may comprise many small groups, e.g., when small, sexually produced colonies first take up symbionts [23]. Overall, however, colonial cnidarians have a population structure that appears entirely unfavorable in this respect, perhaps best characterized as relatively few very large and very long-lasting groups. In other words, colonies are large, long-lived, and relatively scarce, as compared to, say, insects or nematodes, and a single colony contains many trillions of symbionts. Under these conditions, defecting symbionts are strongly selected for within a colony, and additional mechanisms of conflict mediation are a necessity.

Housing symbionts within host cells is a way to create many small groups within a single colony. As described in more detail below, maintaining symbionts in small groups within cells allows other mechanisms to mediate conflict (Fig. 8.1). Replication of symbionts is typically limited by the host, likely by limiting provisioning with inorganic nutrients [22]. Combined with limited replication of symbionts, the biophysics of chemiosmosis dictate that excess product must be exported by the symbiont, i.e., shared with the higher-level unit. If the transporter pathways in the symbiont are inactivated by mutation, the redox state of the now-defecting symbiont will shift in the direction of reduction because of product inhibition or other mechanisms [23]. High levels of reactive oxygen species will form, in turn triggering programmed cell death.

This hypothesis can illuminate features of the coral-dinoflagellate natural history that otherwise are difficult to explain. Symbiotic corals often release large quantities of reduced carbon [24, 25]. Much of this substrate is released in the form of mucus, which includes lipids and polysaccharides and is utilized by many other organisms [24]. Coral workers have struggled to explain how this is adaptive; it may, however, simply be a mechanism to disperse excess product into the environment. Indeed, as pointed out by Crossland et al. [24]: "Lipid production may provide an alternative to zooxanthellar photorespiratory processes…in utilizing excess photosynthetically produced ATP and reducing power [NAD(P)H]….A variety of mechanisms for dispersal of reducing power may be an important feature in maintaining chloroplastic integrity of zooxanthellae contained by the sessile coral in high light environments (e.g., reef flats, shallow reefs)." As was commonplace at the time, Crossland et al. [24] refer to the Symbiodiniaceae as zooxanthellae.

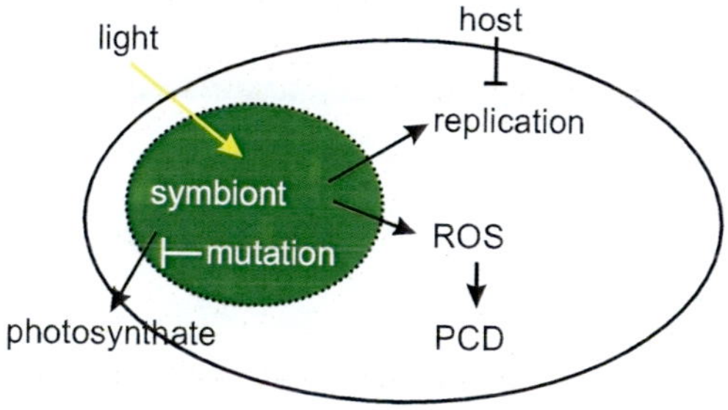

Fig. 8.1 Schematic of evolutionary conflict and its mediation in cnidarian photosymbioses. Via mechanisms that are only partly understood, replication of the symbiont is constrained by the host cell. To avoid product inhibition, the symbiont usually exports reduced carbon. Various mutations to transporter pathway genes can inactivate this export. In the presence of light, product inhibition and other mechanisms result in reactive oxygen species (ROS) formation, which in turn trigger programmed cell death (PCD) of the host cell. (From Blackstone [14])

Generally, as pointed out in Chap. 2, oxygenic photosynthesis is a fraught process, particularly if there is a backup of electrons and the electron carriers become reduced [26]. Production of ROS is greatly enhanced under these conditions [27], and ROS can lead to programmed cell death [28, 29]. If the rate of CO_2 fixation slows with the accumulation of photosynthate, and water splitting proceeds apace, then electrons will back up on photosystems I and II and other electron carriers, ROS formation will increase, and individual symbionts risk serious damage and death. Indeed, such end-product inhibition is a recurring theme in the literature of plant photosynthesis [7, 30, 31], and may be relevant to corals and their symbionts as Crossland et al. [24] point out. With their growth and reproduction limited, symbionts must export photosynthate for their own survival. Those symbionts that have lost by mutation the capacity to translocate photosynthate may thus sow the seeds of their own destruction. Adaptations of corals to expose symbionts to optimal light levels for photosynthesis [32, 33] are thus a two-edged sword—symbiont photosynthesis can occur at a rapid pace, but symbionts that fail to share the bounty succumb all the more quickly. The effectiveness of end-product inhibition, however, diminishes in proportion to the replication rate of the symbionts. In other words, the process of symbiont replication if allowed to proceed rapidly can consume all available photosynthate [23, 34].

Carbon-concentrating mechanisms (CCMs) provide a related mechanism of conflict mediation, focused not on the biochemical products of photosynthesis, but on the substrate. Corals contain CCMs that serve to take up HCO_3^- from seawater and deliver it as CO_2 to symbiont-containing host cells [35–37]. Recently, it has been suggested that individual symbiont-containing host cells rely on energy derived from the oxidation of photosynthates to power their CCMs [35, 36]. In brief, within a host cell, symbiont-produced photosynthate is transported to host coral mitochondria where it is oxidized to form ATP, which powers the host CCMs that are responsible for providing symbionts with an adequate supply of CO_2. As Wooldridge [35, 36] points out, a symbiont that has mutationally lost the ability to translocate photosynthates will quickly impair the ATP supply to its associated host CCMs. In this way, the defecting symbiont will effectively limit its own CO_2 supply, with the same detrimental consequences as end-product inhibition. While similar, CCM collapse may be more effective than the latter process because it does not require a constraint on replication to be effective. CCM collapse, however, does require that the symbiont reside inside a host cell that has functional mitochondria and transporters and not directly contact seawater, as Symbiodiniaceae have bicarbonate transporters of their own [38, 39].

While all mechanisms of conflict mediation likely have a role in maintaining the coral-dinoflagellate symbioses [23], here the focus is on the ones that have been coopted from features of photosynthesis. These mechanisms are particularly crucial because in the presence of light they can distinguish between cooperating and defecting symbionts. Within the coral, symbiont replication is suppressed, but some replication nevertheless occurs. Replication inevitably leads to mutational variation [40, 41], and typically loss-of-function variants will be common. While many loss-of-function variants will be uniformly selected against, those that inactivate mechanisms of photosynthate translocation can potentially be favored at the level of the symbiont. Given the constraints on replication, defecting symbionts may only gain a slight advantage in replication, but even a small advantage translates into large gains over many symbiont generations in a long-lived coral. On the other hand, the multiplication of these defectors represents an existential threat to the higher-level community. How can these defectors be controlled?

As defectors fail to export photosynthate, both product inhibition and the collapse of CCMs result in the backup of electrons onto photosystems I and II and other electron carriers. As the metabolic state of the electron carriers shifts in the direction of reduction, molecular oxygen is likely to gain electrons, forming ROS. Indeed, "Mehler reactions" frequently involve ROS formed by photosystem I and converted by antioxidants into fully reduced oxygen, i.e., water [28, 29]. ROS above the amounts that can be reduced by antioxidants, however, can trigger damage and programmed cell death [28, 29, 42, 43]. Under environmental conditions favorable to symbiont photosynthesis, a fairly constant background level of cell death in this way likely eliminates those host cells containing defectors (Fig. 8.1). In this way, conflict is mediated in favor of the higher-level community, i.e., the cooperative symbionts and the host.

On the other hand, under stressful environmental conditions, typically excess heat and light, in which normal processes of symbiont photosynthesis fail, many more symbionts become highly reduced and emit high levels of ROS [26–28]. During the initiation of bleaching, end-product inhibition is highly unlikely, while CCM collapse may occur [35, 36]. Alternatively, bleaching may initiate with other mechanisms that trigger reduction and ROS. For instance, the initiating steps in bleaching may involve RuBisCO or other Calvin cycle enzymes [44, 45]. Indeed, studies of different coral species and symbiont taxa suggest there may be a diversity of initiating mechanisms [46]. Nevertheless, as long as the initiating mechanisms trigger reduction and ROS, cooperators will become indistinguishable from defectors, and higher levels of programmed cell death will ensue (Fig. 8.2). Many cooperators may thus be destroyed as suggested by the expelled symbionts in the coral *Galaxea fascicularis* and other scleractinians [47, 48]. In these taxa under normal conditions, expelled symbionts were usually functionally impaired, while this was not the case under bleaching conditions.

Indeed, this hypothesis makes several testable predictions. In particular, in unstressed colonies, do stress-resistant symbionts emit more ROS? In a recent study [49], stress-free colonies of *Sarcothelia* sp. produce greater amounts of ROS than those of *Sympodium* sp. These results were predicted based on the widely studied differences between the symbiont types. Symbionts of *Durusdinium* sp. (found in *Sarcothelia* sp.) seem to be more resistant to stress, while those of *Cladocopium* sp. (found in *Sympodium* sp.) are more productive under stress-free circumstances [21]. Evolutionary conflict may underlie these differences. Because of greater conflict under stress-free conditions, symbionts of *Durusdinium* sp. are hypothesized to release less reduced carbon and produce more ROS. Colonies of *Sarcothelia* sp. putatively must have mechanisms to cope with these ROS. On the other hand, *Cladocopium* sp. are hypothesized to be more cooperative under stress-free

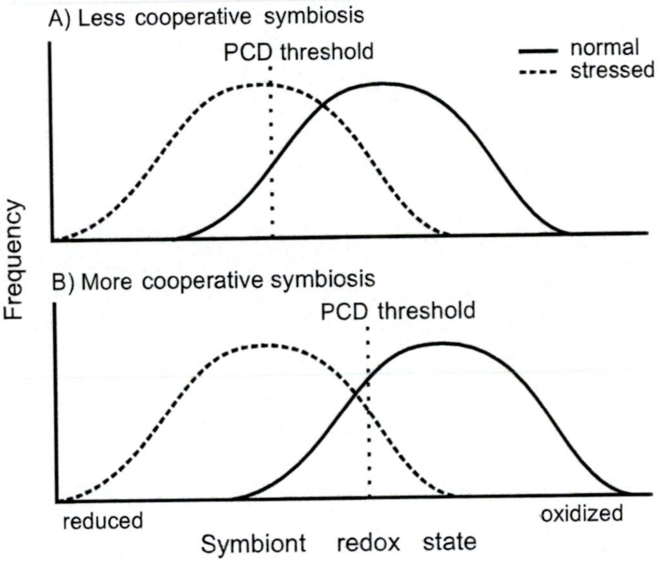

Fig. 8.2 Hypothetical frequency distributions of symbiont redox state (from reduced to oxidized) are shown for (**a**) less and (**b**) more cooperative symbioses in the presence of light. The latter symbiosis better supplies substrate (e.g., CO_2) and removes photosynthate, so symbiont redox state is shifted toward oxidation resulting in low levels of reactive oxygen species. Under normal conditions, in (**a**) only highly reduced symbionts are destroyed by programmed cell death (PCD), while in (**b**) the threshold is higher because of lower investment in antioxidants. Under stressful conditions, photosynthesis breaks down, and the advantages in (**b**) are lost. Both distributions then shift toward a similar relatively reduced distribution, but investment in antioxidants remains the same as do the thresholds for cell death. Symbiont mortality and bleaching is thus greater in (**b**) than in (**a**). (From Blackstone and Golladay [23])

conditions, releasing more reduced carbon and producing less ROS. Colonies of *Sympodium* sp. thus putatively have fewer antioxidant resources to cope with ROS produced during the stress response. What doesn't kill you makes you stronger—colonies conditioned to ROS under normal conditions should thus be more tolerant of them during bleaching. Nevertheless, this between-species comparison should be considered only an exemplar of the sort of results that may be obtained by broader comparisons. Since individual species can have any number of uniquely derived character states, a number of additional comparisons of *Cladocopium*- and *Durusdinium*-containing species must be tested before this hypothesis can be considered well supported.

In summary, high levels of environmental stress cause photosynthesis to break down, and many symbionts may emit high levels of reactive oxygen species, leading to bleaching [22, 27, 28]. Bleaching is a by-product of failed conflict mediation, which leads to the failure of the higher-level unit. Nevertheless, under the same circumstances, Symbiodiniaceae themselves do not bleach. Dinoflagellates are a product of a secondary symbiosis between two eukaryotic cells, one of which contained chloroplasts [50]. This symbiosis is likely based on chemiosmosis as well, but it does not break down, arguably because it exhibits more effective mechanisms of conflict mediation (e.g., genome loss) that entirely prevent this. Adaptation to environmental stress in colonial cnidarians with symbiotic dinoflagellates may require the evolution of more robust mechanisms of conflict mediation.

Lichens By definition, lichens include a variety of terrestrial fungi that form symbiotic associations, usually with cyanobacteria or green algae and sometimes including another fungal partner [51]. These associations may in many ways parallel those of dinoflagellates and marine metazoans, but with some noteworthy differences. The symbionts are generally not housed in cells but rather in a network of hyphae. The symbionts are taxonomically diverse, and it is not clear whether some symbiont types are capable of a free-living existence. In parallel to coral-dinoflagellate symbioses, the benefits of the associations include allowing the algal or bacterial symbionts to achieve the larger size and thus better compete in terrestrial systems, while the host can achieve autotrophy at least in part via symbiont-released carbohydrates. In contrast to dinoflagellates, this apparently occurs passively via a permeabilized cell wall. There are some indications that the fungal host forms structures, called haustoria, which allow extraction of reduced carbon from the symbionts as well as perhaps provisioning them with water and inorganic nutrients [51].

Given the extracellular location of the symbionts, mechanisms of conflict mediation in lichens are unlikely to be based on chemiosmosis, as are those in dinoflagellate-animal symbioses. Rather, it may be that conflict arises when symbiont mutations counteract permeabilization of the cell wall, while mediation occurs simply by the host breaking down and assimilating symbionts that cease to export reduced carbon.

Plant-Rhizobia Symbioses In contrast to the previous examples, the host is the photosynthetic partner in these relationships. Provisioning of soil bacteria could thus proceed with little cost, and, in the case of rhizobia, a considerable gain in the form of nitrogen fixation [52]. Nevertheless, the provisioning of the host with fixed nitrogen does not involve a chemiosmotic process, so it cannot be facilitated, nor evolutionary conflict mediated, by chemiosmosis. Possibly, conflict is mediated by general defenses of plants against parasites, e.g., the hypersensitive response [53].

Plant-Arbuscular Mycorrhizal Fungi Paralleling the previous example, the symbiosis between most land plants and fungi of the phylum Glomeromycota is one of the most consequential for terrestrial ecosystems in general and for cultivated crops in particular [54, 55]. Indeed, this symbiosis may have been crucial for the migration of plants to a terrestrial existence [56]. As with lichens, the fungal partner supplies inorganic nutrients and water, while the plant supplies reduced carbon. Further, the benefit to the plant increases with available light [57], suggesting that when conditions are favorable for chemiosmosis, the fungal network may serve as a sink for excess photosynthate.

Sap-Feeding Insects As with rhizobia and other soil bacteria and fungi, sap-feeding insects can be thought of as utilizing the photosynthetic bounty of terrestrial plants, which may be dispersed at little cost. Thus, these seemly parasitic symbionts may be less harmful than they seem. With a diet of nutrient-poor sap, however, these insects typically require endosymbiotic bacteria, which supply crucial nutrients to their hosts. While these bacteria have been compared to chloroplasts and mitochondria [58], there is at least one crucial difference: it is the host insect that is supplying the chemiosmotic products, albeit secondhand, to the bacteria. Thus, as elaborated below, these endosymbiotic bacteria parallel, for example, rhizobia much more than true eukaryotic organelles. Symbiont chemiosmosis cannot facilitate the formation of, nor subsequently mediate conflict in, these symbioses.

Termite-Flagellate Symbioses While there is considerable complexity to these interactions [59], at least in part they involve gut microbes fermenting various ingested materials anaerobically into reduced carbon molecules that can be taken up by the host and utilized by mitochondria. Thus, these sorts of symbioses appear to be merely a step in the digestion of food. As elaborated below, chemiosmosis does not drive the association, nor can it reasonably be expected to mediate conflicts. Conflict may be mediated by digestion or excretion of a microbial community that likely includes a heavy burden of defectors.

What Conclusions Can Be Drawn from the Role of Chemiosmosis in Modern Symbiosis?

To recapitulate the central argument, the biophysics of chemiosmosis, whether of a presumptive host or symbiont, can favor dispersing excess product into the environment. Quantum electron transfer and supercomplex formation drive chemiosmotic processes at extremely high rates, rapidly producing large quantities of products. While chemiosmosis can be modulated by several mechanisms, releasing excess product provides an alternative means to protect against end-product inhibition and a buildup of dangerous by-products such as reactive oxygen species. This “no-cost” sharing—the free lunch you are forced to make—facilitates interspecific groups, and such groups can lead to cooperative symbioses. Even after such groups have formed, however, cooperators are always vulnerable to exploitation by defectors, so additional mechanisms of conflict mediation are usually necessary. Under some, but not all, circumstances, chemiosmosis may further mediate the conflict.

The examples briefly summarized above can be used to delineate the likely circumstances under which chemiosmosis can both initiate associations as well as subsequently provide additional conflict mediation. For instance, in the case of the termite-flagellate symbiosis, and perhaps other gut microbiota, circumstances seem unlikely to encourage either form of cooperation. Consider a group of flagellates inhabiting a termite gut. Assuming these protists anaerobically ferment complex polysaccharides and emit small carbon molecules as waste, the flagellate group as a whole may accrue some increased rates of reaction by the host termite taking up their carbon waste. The group of flagellates also clearly benefits from the termite providing a habitat. Meanwhile, the aerobic termite respires the waste emitted by the flagellates. Essentially, cooperation emerges because of anaerobic-aerobic complementation, rather than chemiosmosis. Further, the selection favoring flagellate cooperation is at the group level, and group size is likely quite large, so this selection is expected to be weak compared to individual selection. Thus, while one of the partners in this symbiosis uses chemiosmosis to process the substrate shared by the anaerobic symbionts, this does not meet the criteria outlined above for cooperation to blossom. In other words, the chemiosmotic partner is not dispersing product into the environment and thus incentivizing cooperation. Further, since the lower-level units are not chemiosmotic, they are not at a particularly high risk from the side effects of product inhibition. Indeed, gut microbiomes, in general, may be highly vulnerable to defectors, e.g., lower-level individuals that release toxins to gain a competitive advantage, with consequent negative effects on the higher-level unit, that is, the entire group of symbionts and the host. The host may mediate this conflict simply by digesting or excreting symbionts and periodically repopulating the gut.

More likely to foster long-term cooperation are those symbioses in which chemiosmotic products are dispersed into the environment, although there are differences here as well. Some plants disperse reduced carbon photosynthate to soil bacteria and insects. In the former case, some bacteria engage in a mutualistic symbiosis by providing the plant with fixed nitrogen. Nevertheless, there may be dramatic differences in scale between the plants and the microbes that inhabit root nodules. When conditions are favorable for photosynthesis, a macroscopic plant likely produces far more photosynthate than these microbes can utilize. Thus, these microbes may represent a relatively small sink for the plant's reduced carbon. Perhaps more consequential are potentially parasitic sap-feeding insects. If these insects significantly diminish the surfeit of photosynthate experienced at times by the plant, they are perhaps less parasitic than might otherwise be expected. Meanwhile, the insects can provision their symbiotic bacteria and obtain essential nutrients at little cost. Despite parallels in genome reduction, it would be misleading to characterize these symbiotic bacteria as analogous to common eukaryotic organelles. As discussed in Chap. 7, chemiosmosis is perhaps the key feature of the latter. The symbiotic bacteria of sap-feeding insects, on the other hand, would seem more similar to the nitrogen-fixing bacteria of some plants.

The symbiosis between plants and arbuscular mycorrhizal fungi may be the most consequential of the plant-based relationships. Not only do plants receive considerable provisioning from the fungal partner, but they reciprocate with up to one-quarter of their photosynthetic products [55]. Likely, the benefit to the plant partner increases with available light [57]. Possibly, at high light levels, the plant avoids end-product inhibition of chemiosmosis by dispersing large quantities of photosynthate at little cost. Indeed, under circumstances favorable to photosynthesis, plants that do not share reduced carbon may inflict a cost on themselves.

In lichens, chemiosmosis may well have driven the initial symbiosis. Photosynthetic microbes may have released excess reduced carbon into the soil to avoid end-product inhibition and maintain redox balance. Fungi took up this substrate and evolved to "farm" these microbes, providing them with water, inorganic nutrients, and shelter. Given the biophysics of chemiosmosis, these benefits were obtained at little cost to the symbionts. Sharing, in this case, seems to be passive via a permeabilized cell wall. Defectors, however, could perhaps evolve a more specialized cell wall, capable of taking up water and inorganic nutrients but limiting the release of the products of photosynthesis. If such defectors could replicate freely, they might endanger the mutualistic symbiosis. The fungal partner could mediate this conflict by perhaps evolving ways to re-permeabilize the cell walls of defectors or by limiting the replication of all symbionts, in which case end-product inhibition would punish defectors.

Limitation of replication seems to be a key feature of marine animal symbioses with representatives of the Symbiodiniaceae. As with lichens, the symbionts are photosynthetic and may have dispersed excess reduced carbon into the environment to escape end-product inhibition. As associations with animals formed, some symbionts may have experienced loss-of-function mutations, limiting the active export of substrate. If these defectors could replicate freely, the mutualistic symbiosis would collapse. At least in the corals, it appears that replication of the symbionts is inhibited. Further, they are housed intracellularly, so if the export of photosynthate is limited, reactive oxygen species triggered by end-product inhibition lead to programmed death of the host cell. In this way, chemiosmotic mechanisms can lead not only to the formation of symbiosis but also to the mediation of evolutionary conflicts that subsequently arise.

In general, symbiosis may have a major role in evolution [60]. Perhaps not coincidentally, this analysis suggests that the circumstances surrounding the two most successful symbioses in the history of life—the mitochondrion and the chloroplast—are those that most favor a role for chemiosmosis. Small, chemiosmotic symbionts released products to non-chemiosmotic cells. Again, while the origin of mitochondria and eukaryotes is an ancient event in the history of life, it is plausible that chemiosmosis had a twofold impact: triggering the formation of groups that led to a structured population and mediating conflict in favor of cooperators once defectors arose in these groups.

Chemiosmosis may thus have had a powerful impact on the two most consequential symbioses in the history of life. It may also impact many modern symbioses particularly those that parallel the origin of mitochondria and chloroplasts: chemiosmotic symbionts releasing products to maintain redox homeostasis, thus leading to the formation of groups, and as the symbiosis develops, chemiosmosis being co-opted into further mediating evolutionary conflict. While there are many modern eukaryotic symbioses in which one partner is chemiosmotic, many of these do not fit this paradigm, although some do. The powerful effects of chemiosmosis should likely be considered when drawing evolutionary parallels between modern symbioses and events that occurred earlier in the history of life.

References

1. Bronstein JL (ed) (2015) Mutualism. Oxford University Press, Oxford
2. Blackstone NW (2016) An evolutionary framework for understanding the origin of eukaryotes. Biology 5:18
3. Nowak MA (2006) Five rules for the evolution of cooperation. Science 314:1560–1563
4. Michod RE (2003) Cooperation and conflict mediation during the origin of multicellularity. In: Hammerstein P (ed) Genetic and cultural evolution of cooperation. MIT Press, Cambridge, MA, pp 291–307
5. Radzvilavicius AL, Blackstone NW (2018) The evolution of individuality, revisited. Biol Rev 93:1620–1633
6. Chance B, Williams GR (1956) The respiratory chain and oxidative phosphorylation. Adv Enzymol Relat Subj Biochem 17:65–134
7. Goldschmidt EE, Huber SC (1992) Regulation of photosynthesis by end-product accumulation in leaves of plants storing starch, sucrose, and hexose sugars. Plant Physiol 99:1443–1448
8. Service RF (2019) Renewable bonds. Science 365:1236–1239
9. Allen JF, Santabarbara S, Allen CA, Puthiyaveetil S (2011) Discrete redox signaling pathways regulate photosynthetic light-harvesting and chloroplast gene transcription. PLoS One 6:e26372
10. Malone LA, Qian P, Mayneord GE, Hitchcock A, Farmer DA, Thompson RF, Swainsbury DJK, Ranson NA, Hunter NA, Johnson MP (2019) Cryo-EM structure of the spinach cytochrome b6f complex at 3.6 Å resolution. Nature 575:535–539
11. Chance B, Nishimura M (1960) On the mechanism of chlorophyll-cytochrome interaction: the temperature insensitivity of light-induced cytochrome oxidation in chromatium. Proc Natl Acad Sci U S A 46:19–24
12. Moser CC, Keske JM, Warncke K, Farid RS, Dutton PL (1992) Nature of biological electron transfer. Nature 355:796–802
13. Dudkina NV, Eubel H, Keegstra W, Boekema EJ, Braun H-P (2005) Structure of a mitochondrial supercomplex formed by respiratory-chain complexes I and III. Proc Natl Acad Sci U S A 102:3225–3229
14. Blackstone NW (2020) Chemiosmosis, evolutionary conflict, and eukaryotic symbiosis. In: Kloc M (ed) Symbiosis: cellular, molecular, medical, and evolutionary aspects. Springer, Cham, pp 237–252
15. Lane N (2009) Life ascending: the ten great inventions of evolution. Oxford University Press, Oxford
16. Wulff JL (1985) Clonal organisms and the evolution of mutualism. In: Jackson JBC, Buss LW, Cook RE (eds) Population biology and evolution of clonal organisms. Yale University Press, New Haven, pp 437–466
17. Burgess SC, Ryan WH, Blackstone NW, Edmunds PJ, Hoogenboom MO, Levitan DR, Wulff JL (2017) Metabolic scaling in modular animals. Invertebr Biol 136:456–472
18. Bang C, Dagan T, Deines P, Dubilier N, Duschl WJ, Fraune S, Hentschel U, Hirt H, Hülter N, Lachnit T, Picazo D, Pita L, Pogoreutz C, Rädecker N, Saad MM, Schmitz RA, Schulenburg H, Voolstra CR, Weiland-Bräuer N, Ziegler M, Bosch TCG (2018) Metaorganisms in extreme environments: do microbes play a role in organismal adaptation. Zoology 127:1–19
19. Blackstone NW (2021) Evolutionary conflict and coloniality in animals. J Exp Zool Part B (Mol Dev Evol) 336:212–220. https://doi.org/10.1002/jez.b.22924
20. Hoegh-Guldberg O (1999) Climate change, coral bleaching and the future of the world's coral reefs. Mar Freshw Res 50:839–866
21. LaJeunesse TC, Parkinson JE, Gabrielson PW, Jeong HJ, Reimer JD, Voolstra CR, Santos SR (2018) Systematic revision of Symbiodiniaceae highlights the antiquity and diversity of coral endosymbionts. Curr Biol 28:2570–2580
22. Davy SK, Allemand D, Weis VM (2012) Cell biology of cnidarian-dinoflagellate symbiosis. Microbiol Mol Biol Rev 76:229–261

23. Blackstone NW, Golladay JM (2018) Why do corals bleach? Conflict and conflict mediation in a host-symbiont community. BioEssays 40:1800021
24. Crossland CJ, Barnes DJ, Borowitzka MA (1980) Diurnal lipid and mucus production in the staghorn coral Acropora acuminata. Mar Biol 60:81–90
25. Muscatine L, Falkowski P, Porter J, Dubinsky Z (1984) Fate of photosynthetic-fixed carbon in light and shade-adapted colonies of the symbiotic coral Stylophora pistillata. Proc R Soc Lond B 222:181–202
26. Lutz A, Raina J-B, Motti CA, Miller DJ, van Oppen MJH (2015) Host coenzyme Q redox state is an early biomarker of thermal stress in the coral *Acropora millepora*. PLoS One 10:e0139290
27. Parrin AP, Somova EL, Kern PM, Millet TA, Bross LS, Blackstone NW (2017) The use of in vivo microscopy to image the cnidarian stress response. Invertebr Biol 136:330–344
28. Weis VM (2008) Cellular mechanisms of cnidarian bleaching: stress causes the collapse of symbiosis. J Exp Biol 211:3059–3066
29. Paxton CW, Davy SK, Weis VM (2013) Stress and death of cnidarian host cells play a role in cnidarian bleaching. J Exp Biol 216:2813–2820
30. Azcon-Bieto J (1983) Inhibition of photosynthesis by carbohydrates in wheat leaves. Plant Physiol 73:681–686
31. Sawada S, Kuninaka M, Watanabe K, Sato A, Kawamura H, Komine K, Sakamoto T, Kasai M (2001) The mechanism to suppress photosynthesis through end-product inhibition in single-rooted soybean leaves during acclimation to CO2 enrichment. Plant Cell Physiol 42:1093–1102
32. Anthony KRN, Hoogenboom MO, Connolly SR (2005) Adaptive variation in coral geometry and the optimization of internal colony light climates. Funct Ecol 19:17–26
33. Klaus JS, Budd AF, Heikoop JM, Fouke BW (2007) Environmental controls on corallite morphology in the reef coral *Montastraea annularis*. Mar Sci 28:233–260
34. Baker DM, Freeman CJ, Wong JCY, Fogel ML, Knowlton N (2018) Climate change promotes parasitism in a coral symbiosis. ISME J 12:921–930
35. Wooldridge SA (2010) Is the coral-algae symbiosis really 'mutually beneficial' for the partners? BioEssays 32:615–625
36. Wooldridge SA (2013) Breakdown of the coral-algae symbiosis: towards formalising a linkage between warm-water bleaching thresholds and the growth rate of the intracellular zooxanthellae. Biogeosciences 10:1647–1658
37. Goiran C, Al-Moghrabi S, Allemand D, Jaubert J (1996) Inorganic carbon uptake for photosynthesis by the symbiotic coral/dinoflagellate association I. Photosynthetic performances of symbionts and dependence on sea water bicarbonate. J Exp Mar Biol Ecol 199:207–225
38. Lin S, Chieng S, Song B, Zhong X, Lin X, Li W, Li L, Zhang Y, Zhang H, Ji Z, Cai M, Zhuang Y, Shi X, Lin L, Wang L, Wang Z, Liu X, Yu S, Zeng P, Hao H, Zou Q, Chen C, Li Y, Wang Y, Xu C, Meng S, Xu X, Wang J, Yang H, Campbell D, Sturm NR, Dagenais-Bellefeuille S, Morse D (2015) The *Symbiodinium kawagutii* genome illuminates dinoflagellate gene expression and coral symbiosis. Science 350:691–694
39. Aranda M, Li Y, Liew YJ, Baumgartner S, Simakov O, Wilson MC, Piel J, Ashoor H, Bougouffa S, Bajic VB, Ryu T, Ravasi T, Bayer T, Micklem G, Kim H, Bhak J, LaJeunesse TC, Voolstra CR (2016) Genomes of coral dinoflagellate symbionts highlight evolutionary adaptations conducive to a symbiotic lifestyle. Sci Rep 6:39734
40. Michod RE (1999) Darwinian dynamics. Princeton University Press, Princeton
41. van Oppen MJH, Souter P, Howells EJ, Heyward A, Berkelmans R (2011) Novel genetic diversity through somatic mutations: fuel for adaptation of reef corals? Diversity 3:405–423
42. Downs CA, Kramarsky-Winter E, Martinez J, Kushmaro A, Woodley CM, Loya Y, Ostrander GK (2009) Symbiophagy as a cellular mechanism for coral bleaching. Autophagy 5:211–216
43. Dunn SR, Schnitzler CE, Weis VM (2007) Apoptosis and autophagy as mechanisms of dinoflagellate symbiont release during cnidarian bleaching: every which way you lose. Proc R Soc B 274:3079–3085
44. Lesser MP (1996) Elevated temperatures and ultraviolet radiation cause oxidative stress and inhibit photosynthesis in symbiotic dinoflagellates. Limnol Oceanogr 41:271–283

45. Jones RJ, Hoegh-Guldberg O, Larkum AWD, Schreiber U (1998) Temperature-induced bleaching of corals begins with impairment of the CO_2 fixation mechanism in zooxanthellae. Plant Cell Environ 21:1219–1230
46. Buxton L, Takahashi S, Hill R, Ralph PJ (2012) ARIABILITYVariability in the primary site of photosynthetic damage in Symbiodinium sp. (Dinophyceae) exposed to thermal stress. J Phycol 48:117–126
47. Bhagooli R, Hidaka M (2004) Release of zooxanthellae with intact photosynthetic activity by the coral *Galaxea fascicularis* in response to high temperature stress. Mar Biol 145:329–337
48. Fujise L, Yamashita H, Suzuki G, Koike L (2013) Expulsion of zooxanthellae (*Symbiodinium*) from several species of scleractinian corals: comparison under non-stress conditions and thermal stress conditions. Galax J Coral Reef Stud 15:29–36
49. Strohecker J, Golladay J, Paramo M, Paramo M, El Rahmany W, Blackstone NW (2021) Reactive oxygen species and the stress response in Octocorals. Physiol Biochem Zool (in review)
50. Janouškovec J, Horák A, Oborník M, Lukeš J, Keeling PJ (2010) A common red algal origin of the apicomplexan, dinoflagellate, and heterokont plastids. Proc Natl Acad Sci U S A 107:10949–10954
51. Nash TH III (ed) (2008) Lichen biology, 2nd edn. Cambridge University Press, Cambridge
52. Spaink HP, Kondorosi A, Hooykaas PJJ (1998) The Rhizobiaceae. Kluwer Press, Dordrecht
53. Dangol S, Chen Y, Hwang BK, Jwa N-S (2019) Iron- and reactive oxygen species-dependent ferroptotic cell death in rice-Magnaporthe oryzae interactions. Plant Cell 31:189–209
54. Jacott CN, Murray JD, Ridout CJ (2017) Trade-off in arbuscular mycorrhizal symbiosis: disease resistance, growth responses and perspectives for crop breeding. Agronomy 7:75
55. Wipf D, Krajinski F, van Tuinen D, Recorbet G, Courty P-E (2019) Trading on the arbuscular-mycorrhizal networks. New Phytol 223:1127–1142
56. Bouwmeester HJ (2021) Plant lipids enticed fungi to mutualism. Science 372:789–790
57. Johnson NC, Graham JH, Smith FA (1997) Functioning of mycorrhizal associations along the mutualism-parasitism continuum. New Phytol 135:575–585
58. McCutcheon JP (2016) From microbiology to cell biology: when an intracellular bacterium becomes part of its host cell. Curr Opin Cell Biol 41:132–136
59. Brune A, Friedrich M (2000) Microecology of the termite gut: structure and function on a microscale. Curr Opin Microbiol 3:263–269
60. Martin WF (2017) Physiology, anaerobes, and the origin of mitosing cells 50 years on. J Theor Biol 434:2–10

Chapter 9
The Evolution of Multicellularity

Comparative anatomy is largely the story of the struggle to increase surface in proportion to volume.

J. B. S. Haldane [1]

Because eukaryotes are free of the surface-to-volume constraints associated membrane-bound electron transport chains, they are considerably larger than prokaryotes (Chap. 7). These larger sizes subject aerobic eukaryotes to new constraints related to geometry, particularly those that involve bringing food and oxygen into the cell. Why should size matter in this regard? Oxygen diffuses into the cell and such diffusion is dependent on the surface area of the cell. Food can be brought in via transporters on the surface of the cell or by phagocytosis, and both of these processes depend on the surface area. As with membrane-bound energy conversion, this introduces problems of scale because roughly speaking the volume of a geometric object increases with the cube of a linear dimension, while the surface area increases as the square of a linear dimension (Fig. 9.1). As size increases, surface transport or diffusion thus becomes less and less effective, because there is a relatively smaller surface supplying a larger volume. This will have a variety of consequences for large organisms (e.g., large people who play American football are very sensitive to heat because they depend on their surface for cooling).

Besides football, there are a number of circumstances in which larger size is nevertheless desirable (efficient dispersal, exploitation of more or different food sources, producing more offspring, escaping predators, avoidance of the constraints of low Reynolds numbers) [3]. Reynolds numbers are the ratio of forces of momentum to viscosity. This ratio affects the movement of objects in fluids, e.g., small, slow-moving things can "stop still," unlike the Titanic. In any event, in most ecosystems, there is usually "room at the top," i.e., ecological opportunities at the top of the size scale [4]. Perhaps the only consistent disadvantage of large size other than surface-to-volume constraints is a slower replication rate; at times selection for the various advantages seems to have overcome these two disadvantages. Selective

N. W. Blackstone, *Energy and Evolutionary Conflict*,
https://doi.org/10.1007/978-3-031-06059-5_9

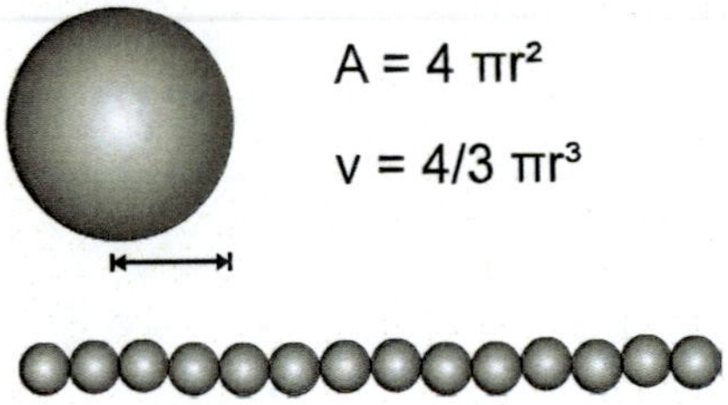

Fig. 9.1 Because surface area (A) increases as the square of the linear dimension, while volume (V) increases as the cube, strings or sheets of small cells escape the surface-to-volume constraints of a single large cell. Multicellularity was one of the first steps in the "struggle to increase surface in proportion to volume" [1]. (From Torday et al. [2])

pressure to increase size on one hand and get substances into cells on the other led to "the struggle to increase surface in proportion to volume" [1]. Multicellular organisms took some of the first steps in this struggle. Many small cells arranged in a chain or a sheet will have a greater surface area relative to volume than a single large cell (Fig. 9.1).

The first radiation of animals and perhaps other kinds of multicellular life occurred sometime in the late Proterozoic. While the "first great oxygen rise" was more than 2200 million years ago, the "second great oxygen rise" occurred at this time, and oxygen levels in the atmosphere may have been approaching those of modern environments [5–10]. The evolution of multicellular life may have been tied to extrinsic oxygen levels. An increase in dissolved oxygen in the ocean may have been caused by increased atmospheric O_2, or by a change in ocean temperature allowing greater oxygen solubility, or both. Related to this, O_2 in the atmosphere may have increased as a consequence of greater burial of organic carbon. Because photosynthesis and respiration typically balance out (Chap. 2), a net surplus of oxygen accumulates only if a corresponding amount of reduced carbon is buried before it can be oxidized. This may have occurred because of greater erosion on land (perhaps in turn caused by the emergence of fungi), or because of the emergence of macroscopic animals with guts in the ocean (macroscopic feces are more likely to be buried before they are consumed by bacteria). Thus, macroscopic life may have exerted positive feedback on O_2 levels by increasing carbon burial. Ultimately, the role of oxygen was that of a terminal electron acceptor in aerobic energy conversion. Increased oxygen levels thus fueled the rise of active animals that were dependent on their mitochondria for efficient utilization of food.

To facilitate exchange with the environment, a multicellular organism can be flat like some of the Ediacaran creatures (or modern placozoans [11] or flatworms) or can exhibit complex branching and folding of the body as in modern sponges. Sponges have the essential design of an internal transport system [12]. In other words, sponges do not rely on simple diffusion of, say, oxygen; they use convection, that is, they ventilate their interiors with a continual supply of the external medium (Fig. 9.2). This essential feature of moving fluid is central to all transport systems. Further elaborations include pumps coupled with efficient mechanics [13, 14]. Oxygen-carrying proteins allow moving less fluid and more oxygen. Nevertheless, with or without these embellishments, internal transport systems share an essential

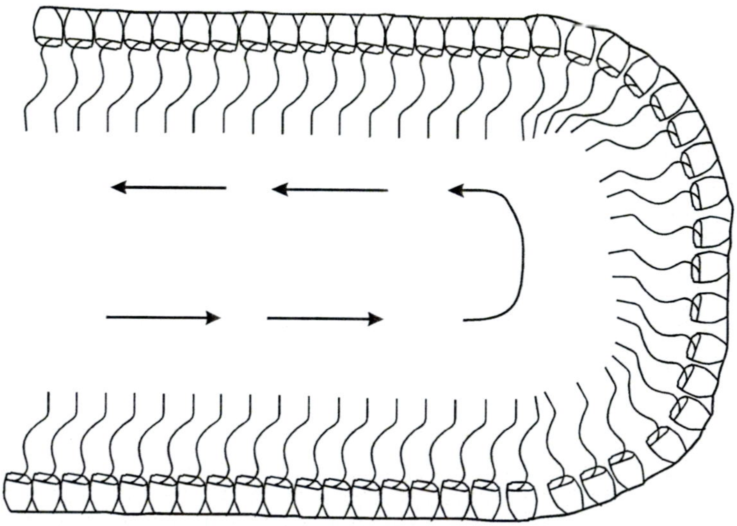

Fig. 9.2 A sketch of the cross section of a small area of the water transport system in a sponge. Monociliated cells move fluid in a coordinated manner through system (arrows). These cells also take up food from the moving fluid (e.g., Berquist [12])

feature: for such a system to work, many somatic cells must become specialized, i.e., dedicated to functions other than the reproduction of the organism. As a consequence, such cells will typically be nondividing.

This can be most clearly understood in the case of ciliated cells lining a simple animal transport system (Fig. 9.2). Animals are opisthokonts, a kind of eukaryote that at least primitively has only one cilium per cell [15]. The structure that allows a eukaryotic cell to replicate is the same structure that produces a cilium. Hence, animal cells cannot bear a cilium and at the same time divide [16]. In practice, this constraint may be circumvented by having groups of ciliated cells work in shifts, e.g., half can divide, while the other half remain ciliated [17]. Nevertheless, cells that were ciliated at least part of the time would divide more slowly than ones that were never ciliated. A multicellular organism with a transport system would therefore tend to favor a division of labor: some cells would carry out cell division to maintain the organism, while other cells would tend to carry out functional chores such as propelling fluid either through or around the organism.

To recapitulate, because of surface-to-volume constraints, larger size entailed multicellularity and ultimately internal transport systems. This in turn required that some cells of the multicellular organism be dedicated to specialized, nonreproductive functions. Thus, in multicellular organisms, there is usually a distinction between somatic cells (specialized cells, which are often nondividing, and which carry out tasks like circulation) and germinal cells (cells which are capable of unlimited division; these cells produce new multicellular individuals).

Herein lies the major peril of multicellularity [16]. Consider an organism evolving from unicellular protists to a multicellular grade. The benefits of multicellularity may be clear, but what are the potential costs? This question will be examined using a choanoflagellate-like protozoan (Fig. 9.3). Such an organism might show a simple multicellular morphology similar to sponges in which ameboid cells in a gelatinous matrix are surrounded by ciliated cells (note that the traditional distinction between flagellum and cilium is not made because both exhibit similar ultrastructure [15]). Such a morphology would facilitate feeding in the bacteria-rich Proterozoic Ocean [18]. The ciliated cells are incapable of cell division or at least replicate more slowly than the ameboid cells. Further, it is the ameboid cells that produce the cells that replicate the entire colony. Ciliated cells are doing the work of moving fluid and taking up food, but these cells are excluded from contributing to the next generation. This is the principal risk of multicellularity. If the ciliated cells are not genetically identical to the ameboid cells, their work in provisioning the multicellular group will be selected against and the colony will fail.

In more formal terms, Darwinian selection at the level of the cell may favor "somatic cell parasites," that is, genetically variant cells, which do not contribute to somatic duties, but rather monopolize reproduction functions. For instance, consider an ameboid cell which, when it divides, does not produce any ciliated cells, only new ameboid cells (Fig. 9.4). Since only ameboid cells divide, the descendants of such a cell will come to predominate in the somatic environment relative to normal ameboid cells, which may produce, say, half ameboid descendants and half ciliated cells. The arithmetic should be clear: the variant ameboid type will outcompete normal cells in the somatic environment. On the other hand, a colony composed of only such selfish variants will quickly degenerate to the unicellular state, with the associated costs. Selection at the level of the cell will favor somatic cell parasites; selection at the level of the multicellular organism will oppose these variants.

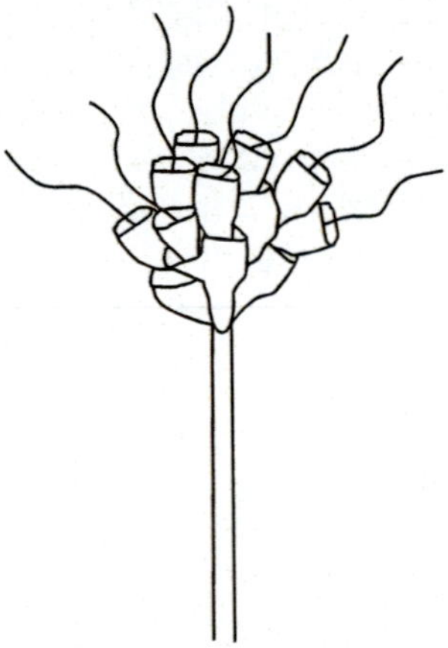

Fig. 9.3 A sketch of a simple multicellular organism, modeled on a modern choanoflagellate. The organism consists of monociliated cells surrounding the hidden ameboid cells in a gelatinous matrix. Ciliated cells move fluid around the colony and trap food, while ameboid cells replicate. The dual nature of life is captured by this division of labor. At the same time, however, ciliated cells are potentially excluded from the next generation. (From Torday et al. [2])

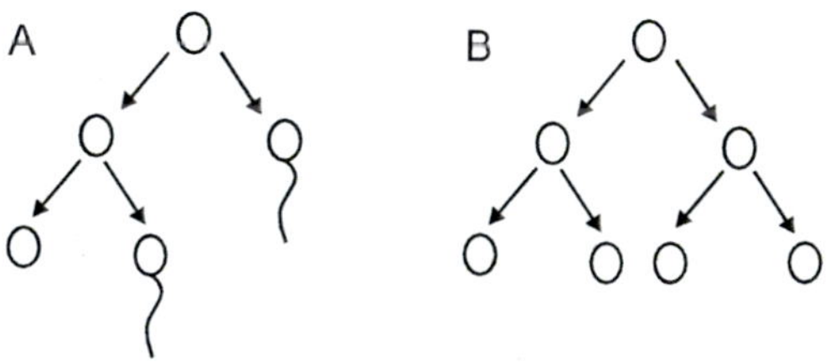

Fig. 9.4 A sketch showing the advantage of a somatic cell parasite. In **a**, with each cell division, a normal ameboid cell produces one somatic cell and one ameboid cell. In **b**, the selfish ameboid cell only produces more of itself. At a low frequency, such selfish cells may successfully parasitize the colony, if somatic functions can be maintained [19]. At a high frequency, such selfish cells will lead to the demise of the entire colony

Multicellular organisms have devised a number of mechanisms to limit such selfish cellular replication [16]. Some of these mechanisms (e.g., a unicellular stage of the life cycle) limit the variation at the lower level and align the evolutionary interests of the cell and the organism via kin selection. Other mechanisms (e.g., programmed cell death) coerce or punish defecting cells and thus also limit the variation at the lower level. Still other mechanisms (sexual reproduction) increase the variation of the higher-level units [19]. Together, these and other mechanisms align selection at the cell and organism level to the extent that the "modern" evolutionary synthesis essentially ignored the former in favor of the latter and still provided a reasonable description of the evolutionary process [16].

In this context, metabolism has been largely ignored. How might metabolism and metabolic gradients mediate evolutionary conflict between the cellular and organismal level? Certainly, as described in Chap. 5, metabolism has been implicated in pattern formation. Further, there is an increasing recognition that proliferative cells share certain metabolic characteristics. Following this brief introduction, in the next chapter, the connections between multicellularity and metabolism are examined in detail.

References

1. Haldane JBS (1927) Possible worlds and other essays. Harper, London
2. Torday JS, Blackstone NW, Rehan VK (2019) Evidence-based evolutionary medicine. Wiley, Hoboken
3. Lachmann M, Blackstone NW, Haig D, Kowald A, Michod RE, Szathmáry E, Werren JH, Wolpert L (2003) Group 3: cooperation and conflict in the evolution of genomes, cells, and multicellular organisms. In: Hammerstein P (ed) Genetic and cultural evolution of cooperation. MIT Press, Cambridge, MA, pp 327–356
4. Bonner JT (1998) The origins of multicellularity. Integr Biol 1:27–36
5. Knoll AH (2003) Life on a young planet. Princeton University Press, Princeton
6. Lane N (2003) Oxygen: the molecule that made the world. Oxford University Press, Oxford
7. Fedonkin MA, Gehling JG, Grey K, Narbonne GM, Vickers-Rich P (2007) The rise of animals. Johns Hopkins University Press, Baltimore

8. Minelli A (2009) Perspectives in animal phylogeny and evolution. Oxford University Press, Oxford
9. Narbonne GM (2010) Ocean chemistry and early animals. Science 328:53–54
10. Li C, Love GD, Lyons TW, Fike DA, Sessions AL, Chu X (2010) A stratified redox model for the Ediacaran ocean. Science 328:80–83
11. Schierwater B (2005) My favorite animal, *Trichoplax adhaerens*. BioEssays 27:1294–1302
12. Bergquist PR (1978) Sponges. University of California Press, Berkeley
13. LaBarbera M (1990) Principles of design of fluid transport systems in zoology. Science 249:992–1000
14. Schmidt-Nielsen K (1997) Animal physiology: adaptation and environment, 5th edn. Cambridge University Press, Cambridge
15. Ax P (1996) Multicellular animals: a new approach to the phylogenetic order in nature, vol 1. Springer, Berlin
16. Buss L (1987) The evolution of individuality. Princeton University Press, Princeton
17. Grosberg RK, Strathmann RR (2007) The evolution of multicellularity: a minor major transition? Annu Rev Ecol Evol Syst 38:621–654
18. Gueneli N, McKenna AM, Ohkouchi N, Boreham CJ, Beghin J, Javaux EJ, Brocks JJ (2018) 1.1-billion-year-old porphyrins establish a marine ecosystem dominated by bacterial primary producers. Proc Natl Acad Sci U S A 115(30):E6978–E6986. https://doi.org/10.1073/pnas.1803866115
19. Michod RE (1999) Darwinian dynamics. Princeton University Press, USA, Princeton

Chapter 10
Metabolism and Multicellularity Revisited

One possible intrinsic difficulty (maybe the biggest hurdle?) is the appropriate down-regulation of cell division at the appropriate time and space in the organism.

Eörs Szathmáry and Lewis Wolpert [1]

Theories relating multicellularity to metabolism date back at least to Charles Manning Child, as described in Chap. 5. Child's ideas lend themselves to modern interpretations [2] in terms of redox gradients and "redox control" [3]. In this context, the dichotomy that confronted Child—does metabolism regulate gene activity or vice versa? —is viewed as a false one. Gene activity and other features of an organism must operate within the constraints of metabolism. In this context, metabolism may be viewed as central to conflict mediation in multicellular organisms. Emerging data from various sources suggest the intriguing possibility that multicellularity involves a shift from continually proliferative "Warburg-like" unicellular organisms to multicellular ones with cell division downregulated in a nutrient-scarce, chemiosmotic somatic environment.

From this perspective, nutrient scarcity may function as a mechanism of conflict mediation, providing the "appropriate downregulation of cell division" by constraining the replication rate of lower-level units [4] and thus the rate of copying errors leading to defecting cells (Fig. 10.1). Much of this decrease may reflect simple consequences of bioenergetics—when nutrients are scarce, replication rates of individual cells are necessarily diminished, decreasing both the costs borne by cooperators and the benefits reaped by defectors. By not cooperating, defectors reap a smaller replicatory reward, while cooperators sacrifice little and may even benefit if the higher-level unit can better harvest the scarce resources that are available. Likely, other stressors can function similarly to nutrient scarcity [5–7].

Certainly, there are numerous examples that suggest a correlation between nutrient scarcity and cooperation. Slime molds and other social microbes typically exist as a unicellular feeding and replicating stage until an area is depleted of nutrients [8]. Cells

N. W. Blackstone, *Energy and Evolutionary Conflict*,
https://doi.org/10.1007/978-3-031-06059-5_10

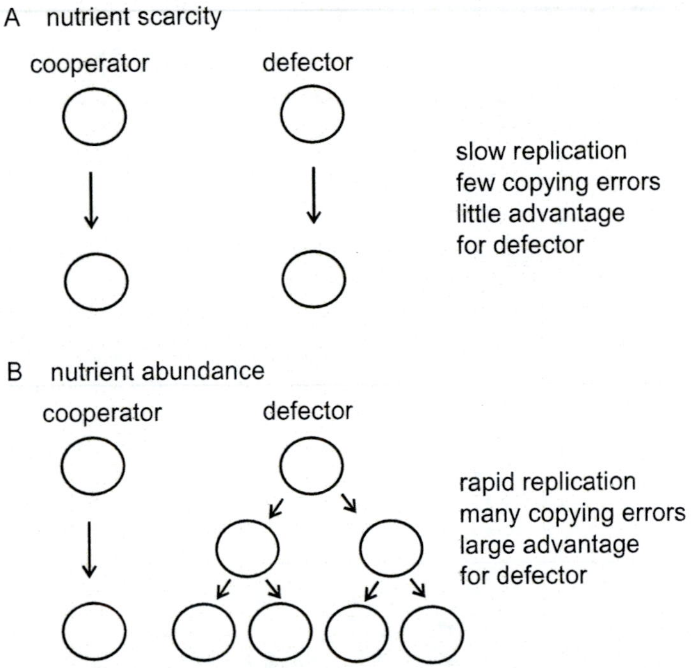

Fig. 10.1 Nutrient scarcity (**a**) mediates conflict in favor of the multicellular organism by constraining the variation at the cell level. By slowing replication, fewer copying errors occur, resulting in a lower mutation rate and less heritable variation. A low replication rate also limits the advantage of a defector cell relative to a cooperator. In contrast, nutrient abundance (**b**) allows fast replication, at least in defectors. A high rate of copying errors increases the mutation rate and heritable variation. Defectors reap a high replicatory reward relative to cooperators. Nutrient scarcity can also mediate conflict by triggering the sexual phase of the life cycle, as, for instance, in slime molds, thus increasing the variance among sexually produced offspring and potentiating selection acting at the higher level. Chemical cytotoxic stress may parallel nutrient scarcity, for example, by triggering growth arrest and limiting the gains of defecting cells. (From Blackstone and Gutterman [7])

then aggregate into a multicellular stage that may carry out the cooperative tasks of movement or reproduction. Greater production of reactive oxygen species (ROS) and activity of antioxidant enzymes occur during the multicellular stages, at least in the slime molds that have been studied [9, 10]. These data suggest that a metabolic shift accompanies the transition from unicellularity to multicellularity, since oxidative phosphorylation will produce more ROS than a mostly fermentative metabolism [11]. This is the case even though the relationship between the rate of oxidative phosphorylation and ROS is not always straightforward [12, 13].

Recent studies by Kelly et al. [14] build upon the earlier work on *Dictyostelium discoideum* and present a somewhat more nuanced view. Nutrient scarcity increased mitochondrial membrane potential and ROS, apparently because of inhibition of the usual path of electron transport. ROS in turn drove the sequestration of sulfur in reduced glutathione, which contains cysteine and is a major antioxidant. Shortages of sulfur then limit mitochondrial metabolism, leading to nonproliferation of cells and multicellular development.

A symbiosis between corals and photosynthetic dinoflagellate symbionts evolved in nutrient-poor waters, forming the basis for coral reef communities. Typically, the coral host limits the provisioning of the symbionts with inorganic nutrients, thus holding the replication of symbionts in check [15, 16] and forcing the symbiont to export photosynthetically produced reduced carbon to the coral, as discussed in Chap. 8. In the presence of excess nutrients, a breakdown of this mutualistic relationship, known as coral bleaching, is more likely to occur [17, 18]. During bleaching, the chemiosmotic process of photosynthesis is usually strongly downregulated within the coral because of the loss or movement of symbionts, and ROS dramatically increase [19–21].

Multinucleate or multicellular fungi have repeatedly evolved into unicellular "yeasts," which are polyphyletic [22]. While the multicellular forms may exhibit greater stress resistance, yeasts usually multiply faster [6], particularly in environments with abundant nutrients [23]. Certainly, some yeasts are well known for their glycolytic metabolism. This suggests a pattern in which unicellular, largely fermentative fungi seem to have repeatedly outcompeted their multicellular predecessors in nutrient-rich environments, while multicellular forms persist in the more challenging environments. The multicellular fungi are predicted to rely on chemiosmosis to a greater extent, although data that explicitly test this prediction are not yet available.

Other stressors may also favor cooperation, at least up to a point. Multicellular organisms are generally more resistant to stress [5, 6], and several well-known symbioses alleviate stress as well, as discussed in Chap. 8. In lichens, for instance, the algal or cyanobacterial symbionts provide photosynthate to the host fungi much like the dinoflagellates in corals. The fungi in turn relieve the water stress of the symbionts and likely also shelter them from the stresses of competition with multicellular plants. In plant-arbuscular mycorrhizal fungi relationships, the host plant is also provisioned with water and inorganic nutrients. Nutrient scarcity and other stressors may trigger generalized stress response pathways in symbiosis and in multicellularity.

The relationship between nutrient scarcity (and perhaps other stressors) and metabolism may have a crucial impact on these natural history patterns. When nutrients are abundant and other stresses are minimal, cells can be profligate with substrate, engaging in glycolysis, which has the benefit of being fast compared to glycolysis plus the Krebs cycle. A fermentative metabolism also has the benefit of providing abundant raw materials for biosynthesis [24–26]. When nutrients are scarce or the cell is under stress, however, the greater yield of chemiosmosis is advantageous, and abundant raw materials for biosynthesis are not required. Chemiosmosis may thus be associated with cooperation in these and other examples, as discussed in previous chapters.

The link between chemiosmosis and cooperation may have a causal basis derived from biophysics. As discussed in Chap. 2, quantum electron transfer and supercomplex formation [27, 28] allow chemiosmosis to rapidly produce considerable amounts of product, raising the risk of end-product inhibition, which is usually termed state 4 metabolism in mitochondria [29]. When chemiosmosis proceeds in the presence of molecular oxygen, end-product inhibition results in the formation of greater amounts of ROS. While several mechanisms can modulate chemiosmosis [30], potential negative effects can also be ameliorated by simply dispersing excess product into the environment. Such largesse can attract individuals of other species leading to groups, in which other organisms share the products that are released into the environment by the chemiosmotic cell or organism. As discussed in Chap. 3, since the time of Darwin, evolutionary biology has recognized that groups are the key to the evolution of cooperation. Groups facilitate kin selection and reciprocity. Further, with many, small groups, chance associations of cooperators can arise, even if cooperation is selected against at the individual level. Groups of cooperators can then outcompete groups of defectors.

As discussed previously, sharing of chemiosmotic products was likely central to the symbiosis that led to the eukaryotic cell [31]. To transcend the surface-to-volume constraints that limit the size of prokaryotes, energy-converting lower-level units (the protomitochondria) had to export these products to the cytosol [32]. This form of cooperation invites defection based on loss-of-function mutations, and mechanisms of conflict mediation based on chemiosmotic signaling necessarily developed [12, 13]. Eukaryotic signaling associated with a chemiosmotic metabolism may thus enhance cooperation and limit defection. This signaling may have biomedical relevance in a number of contexts, for instance, contributing to the pathways that affect longevity [33]. Prosocial effects may be lost, however, if a cell adopts a largely fermentative metabolism and downregulates chemiosmosis.

These metabolic factors may be related to the breakdown of cooperation in multicellular organisms. Multicellularity has evolved repeatedly in eukaryotes. Each time that it evolved, nutrient sharing was likely one of its foundations [34]. Simple multicellular aggregations may have created physical gradients that led to terminally differentiated ciliated cells surrounding a group of undifferentiated stem cells. This arrangement likely resulted from redox gradients within the colony [3], with differentiated cells on the exterior carrying out oxidative phosphorylation, while stem cells in the low-oxygen interior remained glycolytic and proliferative (Fig. 10.2). Indeed, similar patterns of metabolism and development may be found in modern eukaryotes as well [26].

In a simple colony of cells, interior-to-exterior metabolic gradients provide a level of responsiveness to environmental conditions. If exterior cells obtained ample supplies of food, while interior cells were somewhat starved, there would be an oxidized-to-reduced gradient. On the other hand, if exterior cells obtained sufficient oxygen, while interior cells were somewhat hypoxic, there would be a reduced-to-oxidized gradient. Similarly, if exterior cells were subject to intense metabolic demand, e.g., because of rapid ciliary action, there would also be a reduced-to-oxidized gradient. The direction and intensity of the metabolic gradient could effectively be used to adjust the differentiation of stem cells. A reduced-to-oxidized gradient might signal intense metabolic demand for ciliary action in exterior cells, or insufficient ventilation of interior cells, or both. Differentiating more stem cells into ciliated exterior cells would increase locomotion and ventilation so that the interior-to-exterior metabolic gradient was diminished.

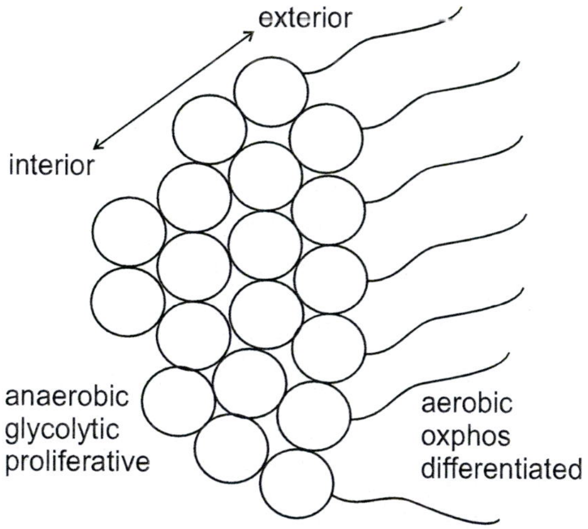

Fig. 10.2 Metabolic gradients are shown in a simple multicellular aggregation. Following a physical gradient from more aerobic to more anaerobic, terminally differentiated ciliated cells on the exterior carry out oxidative phosphorylation (oxphos), while undifferentiated stem cells in the interior remain largely glycolytic and proliferative. (Modified from Blackstone and Gutterman [7])

In every incipient multicellular aggregation, however, defection remained a risk. For the entire history of unicellular life, unfettered replication was selected for, yet multicellularity depends on the appropriate downregulation of cell division [1]. In a multicellular aggregation, a cell may have signaled defection by a fermentative metabolism, nutrient monopolization, and perhaps even cannibalism of its neighboring cells, followed by rapid replication. Cooperator cells likely developed mechanisms to detect and avoid the defector cells. Without robust mechanisms of conflict mediation, simple multicellular organisms may have failed to emerge and reverted to the unicellular state. Even when a multicellular stage was successfully incorporated into the life cycle, cell-level selection has always remained a potent force. Indeed, cancer-like phenomena are found throughout all groups of modern multicellular eukaryotes [34], despite numerous mechanisms of conflict mediation. These mechanisms likely include a unicellular stage of the life cycle, programmed cell death, cell walls, a germ line, and others [35]. Nevertheless, cancers still form with distressing regularity. Modern cancer cells defect from somatic duties and engage in rapid replication. In many ways, their behavior suggests atavistic unicellular organisms [36–38], exhibiting striking parallels to yeasts—polyphyletic, unicellular, fast-replicating, and predominately glycolytic. While the recognition of unique features of cancer metabolism has considerable historical precedent (e.g., the Warburg effect), a much more complete understanding of these features has only recently been developed. Since cancers are governed by cell-level selection, aerobic glycolysis is adaptive (Fig. 10.3). Chemiosmosis and perhaps ROS are downregulated, and biosynthesis upregulated [24, 39, 40]. Cancer cells abrogate the nutrient scarcity that is otherwise enforced in the somatic environment [25].

The glycolytic tendencies of cancer cells lead to preferences for certain nutrients, for instance, metabolizing glucose to generate ATP and regenerating NAD+ by converting pyruvate to lactate. Complementing this metabolism, many cancers use

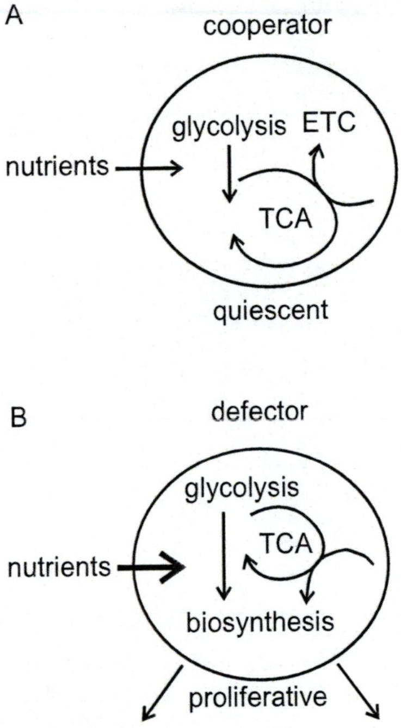

Fig. 10.3 Cooperator cells (**a**) remain terminally differentiated, with low nutrient demands, using the Krebs or tricarboxylic acid cycle (TCA) to reduce cofactors that are oxidized using chemiosmosis in the electron transport chain (ETC), producing ATP. Defector cells (**b**) are proliferative, with high nutrient demands, using glycolysis to produce ATP via substrate-level phosphorylation and, along with the TCA, to generate precursors for biosynthesis. (Modified from Blackstone and Gutterman [7])

glutamine to fuel the Krebs or tricarboxylic acid (TCA) cycle, generating building blocks for biosynthesis and rapid growth [26]. Leone et al. [41] point out that these features (i) parallel those of T cells undergoing rapid proliferation, and (ii) create a tumor microenvironment hostile to an immune response. They hypothesize that inhibiting glutamine metabolism may have therapeutic benefits. Using the glutamine antagonist 6-diazo-5-oxo-L-norleucine (DON), modified to activate only in the tumor microenvironment, they were able to severely disrupt the Warburg metabolism and enhance the effects of other therapeutics, for instance, leading to an appropriate immune response. Indeed, inhibiting glutamine metabolism also had profound effects on T cells, including upregulating oxidative phosphorylation. These remarkable effects, essentially inhibiting defectors and enhancing cooperators, suggest considerable clinical potential.

Kanarek et al. [42] provide a complementary analysis. Many tumors have relatively specific nutrient requirements (e.g., glucose and glutamine), and dietary limitation of consumption of these nutrients can potentiate the effects of therapeutics acting on specific molecular targets. General approaches to dietary manipulations are also possible. For instance, calorie restriction delays the onset of aging-related disease and extends longevity [33, 43]. In unicellular eukaryotes, the effects of calorie restriction may derive from the effects of nutrient scarcity described in the previous section (e.g., favoring a chemiosmotic metabolism rather than a glycolytic one), while in multicellular organisms, calorie restriction's effects may in part depend on a relationship between nutrient scarcity and cancer. The ketogenic diet, developed to treat epilepsy [44], may potentially limit cancer, at least in part by its effects on dietary glucose. Likewise, metformin, used to treat diabetes, has a variety of effects related to downregulating metabolism and may in some cases be a cancer therapeutic [45, 46].

Tumor suppression may also depend on underlying metabolic subtleties that are only beginning to be understood. For instance, TP53 is mutated in most human cancers. Utilizing a mouse model, Morris et al. [47] show that restoring p53 function in cancer cells leads to the accumulation of α-ketoglutarate (αKG), a TCA cycle intermediate. αKG then activates chromatin-modifying enzymes and transcriptional programs that are characteristic of premalignant differentiation. At least in part, exogenous αKG mimics these effects. The data suggest that p53 triggers a metabolic switch, increasing the incorporation of glucose-derived carbons and diminishing the contribution of glutamine-derived carbons into TCA cycle intermediates. Cooperation is thus enhanced, and defection diminished. From an evolutionary standpoint, one might surmise that the metabolic signaling preceded the recruitment of p53 to its current function. In other words, chemiosmosis and glucose-derived carbon in TCA cycle intermediates may have been the original signal for cooperation. Interestingly, αKG and the αKG/succinate ratio may have general roles in differentiation, development, and lifespan [48].

While multicellular organisms seem adept at manufacturing nutrient scarcity in the somatic environment [25], ultimately environmental inputs cannot be entirely overcome. Considerable data thus suggest that nutrient abundance and concomitant effects such as obesity and diabetes are major risk factors for cancer [39, 49]. For instance, Chung et al. [50] focus on pancreatic ductal adenocarcinoma (PDAC), one of the most lethal cancers. KRAS mutations are implicated in the vast majority of these. Inhibition of these mutants is challenging, and PDAC cells seem able to survive the ablation of KRAS, suggesting that inhibition, when it is available, will quickly lead to resistance. As Chung et al. [50] point out: "Therefore, alternative paradigms beyond genetic factors need to be explored to develop novel therapeutic and preventative strategies for PDAC." Indeed, the human body mass index is associated with PDAC risk, and greater values correlate with poorer prognoses. Rather than additional driver mutations, Chung et al. implicate microenvironmental factors, leading to aberrant hormonal expression that promotes KRAS-driven PDAC. Interestingly, in mouse models, nutrient scarcity and weight loss alleviate cancer development.

Both obesity and diabetes are often associated with hyperinsulinemia, which is characterized by high levels of circulating insulin. Using a fly epithelium model, Sanaki et al. [51] show that under nutrient scarcity and normal insulin levels, wild-type cells outcompete premalignant ones. Hyperinsulinemia, which is often induced by high nutrient levels, reverses the competitive relationship among these cells. With greater insulin-mTOR signaling, premalignant cells increase protein synthesis and overgrow wild-type cells, leading to tumorigenesis. Much like the results of Leone et al. [41], these results indicate that nutrient scarcity can inhibit defectors and enhance cooperators.

As discussed above [47], gene activity may reinforce metabolic alterations. Lee et al. [52] target BACH1, a hem-binding transcription factor that has increased expression in tumors from patients with triple negative breast cancer. While increasing aerobic glycolysis, BACH1 decreases glucose utilization in the Krebs or tricarboxylic acid cycle and negatively regulates transcription of electron transport chain genes. Downregulating BACH1 restores the sensitivity of cells to metformin, which among other functions inhibits complex I of the electron transport chain. Limiting expression of BACH1 also suppresses the growth of both cell line and patient-derived tumor xenografts. Again, the implication is that cancer thrives in an environment in which oxidative phosphorylation is downregulated relative to glycolysis.

In this context, considerations of metastasis should be included. In the absence of metastasis, cancers remain relatively treatable. Certainly, multicellular organisms with cell walls are protected against cancers becoming systemic. On the other hand, once a mammalian cancer has metastasized, therapy becomes much more challenging. The initiating step in metastasis is of course the movement of cancer cells. Compared to somatic cells, cancer cells show a greater tendency to migrate [53]. Any sort of stress, including nutrient scarcity, might be expected to accelerate this migration, with negative effects on the patient [54]. Deliberately stressing cancers may thus be called into question on these grounds.

Certainly, many organisms respond to stress in general and nutrient scarcity in particular by dispersing in space, or time, or both. There are, however, crucial differences. Typically, stress or nutrient scarcity triggers the sexual phase of the life cycle, creating genetic diversity of propagules, perhaps followed by the formation of resistant spores or a dedicated dispersal stage. In other words, stress triggers a specialized phase of the life cycle that arguably evolved precisely to allow dispersal when stressful circumstances were encountered.

A crucial weakness of cancer cells is thus apparent. While some features of somatic cells may parallel organismal life cycles and life histories [55], these cells do not in any sense have a true life cycle. For instance, somatic cells are by definition incapable of meiosis, nor do they have elaborate adaptations for dispersal or dormancy. Thus, cancer cells disperse in their usual cellular forms, although perhaps forming small aggregations. Under these circumstances, the success of migrating cancer cells is expected to be low.

Indeed, this is the case. Cancer cells migrate readily but are usually unsuccessful in completing metastasis. Metastasis requires not only migration, but successful progression through several other steps that are likely rate-limiting [56]. While stress or nutrient scarcity might trigger more migration, the effects of this additional migration may be mitigated if the stress further constrains the steps that are actually rate-limiting in metastasis. Stress that acts systemically to alter the tumor microenvironment may thus have a beneficial effect overall [57].

Considerable opportunities for experimental approaches seem indicated. Nevertheless, the conditions that affect nutrient scarcity can be nuanced. Time- or nutrient-restricted diets can have divergent effects [58, 59]. Ulgherait et al. [58] develop protocols for intermittent time-restrictive feeding in *Drosophila* that prolonged lifespan and slowed aging. A diet that resulted in night-specific induction of autophagy was both necessary and sufficient to extend lifespan. Further, the effects of dietary composition can be subtle. As discussed above, both calorie restriction and a ketogenic diet have been implicating in limiting tumor progression by lowering blood glucose and insulin levels. Using mouse models, however, Lien et al. [59] found greater effects of calorie restriction and trace the difference to lower lipid levels, which cause an imbalance between unsaturated and saturated fatty acids and slow tumor growth.

Notably, the model systems (flies and mice) used in these experiments have limited mitotic potential and hence short lifespans. Of course, this is not the case with all metazoans. Examining an organism with essentially unlimited mitotic potential might be illuminating. Colonial cnidarians provide some useful models in this context [60, 61]. Currently, Wean El Rahmany, a graduate student in my laboratory, is leading a series of experiments with a strain of *Eirene* sp. that is notable for its remarkable proliferation and hardiness. These colonies are able to tolerate and even thrive under the exacting conditions of nutrient deprivation. Examining the response of mutation rate and metabolism to dietary manipulation may provide useful insight into these factors in a purely mitotic system.

In any event, substantial evidence, much of it from biomedicine, supports a connection between metabolism and multicellular cooperation. Many of the metabolic mechanisms utilized by multicellular eukaryotes to mediate cell-cell conflicts may have been co-opted from the previous major transition, the origin of eukaryotes [12]. Several obvious predictions (e.g., commonalities between metabolic signaling pathways in eukaryotes that independently derived multicellularity) may not apply. For instance, with cell walls to mediate conflict, plants are relatively immune to cellular defection and simply may not have implemented extensive metabolic signaling to function in this regard. Nevertheless, the parallels between yeasts and cancer cells noted above are intriguing. In the next chapter, again turning to the biomedical literature, another prediction is elaborated: metabolic signaling is deeply embedded in the biology of animals. Some of this signaling may represent vestiges of mechanisms of conflict mediation reaching back to the origin of eukaryotes. The synthesis presented here potentially provides a framework to rationalize at least some of these seemingly baroque cellular and molecular natural histories that are being elucidated.

References

1. Szathmáry E, Wolpert L (2003) The transition from single cells to multicellularity. In: Hammerstein P (ed) Genetic and cultural evolution of cooperation. MIT Press, Cambridge, MA, pp 271–290
2. Blackstone NW (2006) Charles Manning Child (1869–1954): the past, present, and future of metabolic signaling. J Exp Zool (MDE) 306B:1–7
3. Blackstone NW (2000) Redox control and the evolution of multicellularity. BioEssays 22:947–953

4. Michod RE (2003) Cooperation and conflict mediation during the origin of multicellularity. In: Hammerstein P (ed) Genetic and cultural evolution of cooperation. MIT Press, Cambridge, MA, pp 291–307
5. Bonner JT (1998) The origins of multicellularity. Integr Biol 1:27–36
6. Kuzdzal-Fick JJ, Chen L, Balázsi G (2019) Disadvantages and benefits of evolved unicellularity versus multicellularity in budding yeast. Ecol Evol 9:8509–8523
7. Blackstone NW, Gutterman JU (2021) Can natural selection and druggable targets synergize? Of nutrient scarcity, cancer, and the evolution of cooperation. BioEssays 43:2000160
8. Bonner JT (2009) The social amoebae: the biology of cellular slime molds. Princeton University Press, Princeton, p 144
9. Bloomfield G, Pears C (2003) Superoxide signalling required for multicellular development of *Dictyostelium*. J Cell Sci 116:3387–3397
10. Allen RG, Farmer KJ, Toy PL, Newton RK, Sohal RS, Nations C (1985) Involvement of glutathione in the differentiation of the slime mold *Physarum polycephalum*. Dev Growth Differ 27:615–620
11. Allen BG, Bhatia SK, Anderson CM, Eichenberger-Gilmore JM, Sibenaller ZA, Mapuskar KA, Schoenfeld JD, Buatti JM, Spitz DR, Fath MA (2014) Ketogenic diets as an adjuvant cancer therapy: history and potential mechanism. Redox Biol 2:963–970
12. Blackstone NW (2013) Why did eukaryotes evolve only once? Genetic and energetic aspects of conflict and conflict mediation. Philos Trans R Soc B 368:20120266
13. Blackstone NW (2015) The impact of mitochondrial endosymbiosis on the evolution of calcium signaling. Cell Calcium 57:133–139
14. Kelly B, Carrizo GE, Edwards-Hicks J, Sanin DE, Stanczak MA, Priesnitz C, Flachsmann LJ, Curtis JD, Mittler G, Musa Y, Becker T, Buescher JM, Pearce EL (2021) Sulfur sequestration promotes multicellularity during nutrient limitation. Nature 591:471–476
15. Wooldridge SA (2010) Is the coral-algae symbiosis really 'mutually beneficial' for the partners? BioEssays 32:615–625
16. Davy SK, Allemand D, Weis VM (2012) Cell biology of cnidarian-dinoflagellate symbiosis. Microbiol Mol Biol Rev 76:229–261
17. Cunning R, Baker AC (2013) Excess algal symbionts increase the susceptibility of reef corals to bleaching. Nat Clim Change 3:259–262
18. Vega Thurber RL, Burkepile DE, Fuchs C, Shantz AA, McMinds R, Zaneveld JR (2014) Chronic nutrient enrichment increases prevalence and severity of coral disease and bleaching. Glob Change Biol 20:544–554
19. Netherton SE, Scheer DM, Morrison PR, Parrin AP, Blackstone NW (2014) Physiological correlates of symbiont migration during bleaching of two octocoral species. J Exp Biol 217:1469–1477
20. Parrin AP, Somova EL, Kern PM, Millet TA, Bross LS, Blackstone NW (2017) The use of in vivo microscopy to image the cnidarian stress response. Invertebr Biol 136:330–344
21. Almegbel MNA, Rowe EA, Alnaser FN, Yaeger M, Blackstone NW (2019) Metabolic activation and scaling in two species of colonial cnidarians. Biol Bull 237:63–72
22. Nagy LG, Ohm RA, Kovacs GM, Floudas D, Riley R, Gacser A, Sipiczki M, Davis JM, Doty SL, de Hoog GS, Lang BF, Spatafora JW, Martin FM, Grigoriev IV, Hibbett DS (2014) Latent homology and convergent regulatory evolution underlies the repeated emergence of yeasts. Nat Commun 5:1–8
23. Nagy LG, Tóth R, Kiss E, Slot J, Gácser A, Kovács GM (2017) Six key traits of fungi: their evolutionary origins and genetic bases. Microbiol Spectr 5:1–22
24. VanderHeiden MG, Cantley LC, Thompson CB (2009) Understanding the Warburg effect: the metabolic requirements of cell proliferation. Science 324:1029–1033
25. Pavlova NN, Thompson CB (2016) The emerging hallmarks of cancer metabolism. Cell Metab 23:27–47
26. Ly CH, Lynch GS, Ryall JG (2020) A metabolic roadmap for somatic stem cell fate. Cell Metab 31:1052–1067

27. Moser CC, Keske JM, Warncke K, Farid RS, Dutton PL (1992) Nature of biological electron transfer. Nature 355:796–802
28. Dudkina NV, Eubel H, Keegstra W, Boekema EJ, Braun H-P (2005) Structure of a mitochondrial supercomplex formed by respiratory-chain complexes I and III. Proc Natl Acad Sci U S A 102:3225–3229
29. Chance B, Williams GR (1956) The respiratory chain and oxidative phosphorylation. Adv Enzymol Relat Subj Biochem 17:65–134
30. Allen JF, Santabarbara S, Allen CA, Puthiyaveetil S (2011) Discrete redox signaling pathways regulate photosynthetic light-harvesting and chloroplast gene transcription. PLoS One 6:e26372
31. Martin WF, Garg SG, Zimorski V (2015) Endosymbiotic theories for eukaryotic origin. Philos Trans R Soc Lond B 370:20140330
32. Lane N (2005) Power, sex, suicide: mitochondria and the meaning of life. Oxford University Press, Oxford
33. Fontana L, Partridge L (2015) Promoting health and longevity through diet: from model organisms to humans. Cell 161:106–118
34. Aktipis CA, Boddy AM, Jansen G, Hibner U, Hochberg ME, Maley CC, Wilkinson GS (2015) Cancer across the tree of life: cooperation and cheating in multicellularity. Philos Trans R Soc Lond B 370:20140219
35. Lachmann M, Blackstone NW, Haig D, Kowald A, Michod RE, Szathmáry E, Werren JH, Wolpert L (2003) Group 3: Cooperation and conflict in the evolution of genomes, cells, and multicellular organisms. In: Hammerstein P (ed) Genetic and cultural evolution of cooperation. MIT Press, Cambridge, MA, pp 327–356
36. Russo M, Crisafulli G, Sogari A, Reilly NM, Arena S, Lamba S, Bartolini A, Amodio V, Magrì A, Novara L, Sarotto I, Nagel ZD, Piett CG, Amatu A, Sartore-Bianchi A, Siena S, Bertotti A, Trusolino L, Corigliano M, Gherardi M, Lagomarsino MC, Nicolantonio F, Bardelli A (2019) Adaptive mutability of colorectal cancers in response to targeted therapies. Science 366:1473–1480
37. Cipponi A, Goode DL, Bedo J, McCabe MJ, Pajic M, Croucher DR, Rajal AG, Junankar SR, Saunders DN, Lobachevsky P, Papenfuss AT, Nessem D, Nobis M, Warren SC, Timpson P, Cowley M, Vargas AC, Qiu MR, Generali DG, Keerthikumar S, Nguyen U, Corcoran NM, Long GV, Blay J-Y, Thomas DM (2020) MTOR signaling orchestrates stress-induced mutagenesis, facilitating adaptive evolution in cancer. Science 368:1127–1131
38. Chen H, Lin F, Xing K, He X (2015) The reverse evolution from multicellularity to unicellularity during carcinogenesis. Nat Commun 6:1–9
39. Wellen KE, Thompson CB (2010) Cellular metabolic stress: considering how cells respond to nutrient excess. Mol Cell 40:323–332
40. Coller HA (2014) Is cancer a metabolic disease? Am J Pathol 184:4–17
41. Leone RD, Zhao L, Englert JM, Sun I-M, Oh M-H, Sun I-H, Arwood ML, Bettencourt IA, Patel CH, Wen J, Tam A, Blosser RL, Prchalova E, Alt J, Rais R, Slusher BS, Powell J, JD. (2019) Glutamine blockade induces divergent metabolic programs to overcome tumor immune evasion. Science 366:1013–1021
42. Kanarek N, Petrova B, Sabatini DM (2020) Dietary modifications for enhanced cancer therapy. Nature 579:507–517
43. Sinclair DA (2005) Toward a unified theory of caloric restriction and longevity regulation. Mech Ageing Dev 126:987–1002
44. Lutas A, Yellen G (2013) The ketogenic diet: metabolic influences on brain excitability and epilepsy. Trends Neurosci 36:32–40
45. Coll AP, Chen M, Taskar P, Rimmington D, Patel S, Tadross JA, , Cimino I, Yang M, Welsh P, Virtue S, Goldspink DA, Miedzybrodzka EL, Konopka AR, Esponda RR, Huang JT-J, Tung YCL, Rodriguez-Cuenca S, Tomaz RA, Harding HP, Melvin A, Yeo GSH, Preiss D, Vidal-Puig A, Vallier L, Nair KS, Wareham NJ, Ron D, Gribble FM, Reimann F, Sattar N, Savage DB,

Allan BB, O'Rahilly S. 2020. GDF15 mediates the effects of metformin on body weight and energy balance. Nature, 578:444–448
46. Sahra IB, Marchand-Brustel YL, Tanti J-F, Bost F (2010) Metformin in cancer therapy: a new perspective for an old antidiabetic drug? Mol Cancer Ther 9:1092–1099
47. Morris JP, Yashinskie JJ, Koche R, Chandwani R, Tian S, Chen C-C, Baslan T, Marinkovic ZS, Sánchez-Rivera FJ, Leach SD, Carmona-Fontaine C, Thompson CB, Finley LWS, Lowe SW (2019) α-Ketoglutarate links p53 to cell fate during tumour suppression. Nature 573:595–599
48. Borsa M, Simon AK (2020) Fine-tuning stemness. Science 369:373–374
49. Stone TW, McPherson M, Darlington LG (2018) Obesity and cancer: existing and new hypotheses for a causal connection. EBioMedicine 30:14–28
50. Chung KM, Singh J, Lawres L, Dorans KJ, Garcia C, Burkhardt DB, Robbins R, Bhutkar A, Cardone R, Zhao X, Babic A, Vayrynen SA, Costa AD, Nowak JA, Chang DT, Dunne RF, Hezel AF, Koong AC, Wilhelm JJ, Bellin MD, Nylander V, Gloyn AL, McCarthy MI, Kibbey RG, Krishnaswamy S, Wolpin BM, Jacks T, Fuchs CS, Muzumdar MD (2020) Endocrine-exocrine signaling drives obesity associated pancreatic ductal adenocarcinoma. Cell 181:832–847
51. Sanaki Y, Nagata R, Kizawa D, Leopold P, Igaki T (2020) Hyperinsulinemia drives epithelial tumorigenesis by abrogating cell competition. Dev Cell 53:379–389
52. Lee J, Yesilkanal AE, Wynne JP, Frankenberger C, Liu J, Yan J, Elbaz M, Rabe DC, Rustandy FD, Tiwari P, Grossman EA, Hart PC, Kang C, Sanderson SM, Andrade J, Nomura DK, Bonini MG, Locasale JW, Rosner MR (2019) Effective breast cancer combination therapy targeting BACH1 and mitochondrial metabolism. Nature 568:254–258
53. Aktipis CA, Maley CC, Pepper JW (2012) Dispersal evolution in neoplasms: the role of disregulated metabolism in the evolution of cell motility. Cancer Prev Res 5:266–275
54. García-Jiménez C, Goding CR (2019) Starvation and pseudo-starvation as drivers of cancer metastasis through translation reprogramming. Cell Metab 29:254–267
55. Wang A, Luan H, Medzhitov R (2019) An evolutionary perspective on immunometabolism. Science 363:eaar3932
56. Massagué J, Obenauf AC (2016) Metastatic colonization by circulating tumour cells. Nature 529:298–306
57. Bissell MJ, Hines WC (2011) Why don't we get more cancer? A proposed role of the microenvironment in restraining cancer progression. Nat Med 17:320–329
58. Ulgherait M, Midoun AM, Park SJ, Gatto JA, Tener SJ, Siewert J, Klickstein N, Canman JC, Ja WW, Shirasu-Hiza M (2021) Circadian autophagy drives iTRF-mediated longevity. Nature 598:353–358
59. Lien EC, Westermark AM, Zhang Y, Yuan C, Li Z, Lau AN, Sapp KM, Wolpin BM, Vander Heiden MG (2021) Low glycaemic diets alter lipid metabolism to influence tumour growth. Nature 599:302–307
60. Blackstone NW, Bridge DM (2005) Model systems for environmental signaling. Integr Comp Biol 45:605–614
61. Blackstone NW, Kelly MM, Haridas V, Gutterman JU (2005) Mitochondria as integrators of information in an early-evolving animal: insights from a triterpenoid metabolite. Proc R Soc Lond B 272:527–531

Chapter 11
Metabolic Vestiges of Conflict Mediation in Modern Biology

Eukaryotes are complex and the pivotal role of mitochondria in the origin of that complexity…seems increasingly difficult to dispute….

Bill Martin [1]

Given that life can be characterized by replication and information on one hand and energy conversion on the other, there is every reason to suspect that selection has favored cells and organisms that effectively tune their metabolism to environmental conditions. On the other hand, the perspective suggested here—that the complex interplay between metabolism and evolutionary conflict was central to the history of life—puts a rather different focus on metabolic mechanisms: multicellular eukaryotes should be a veritable museum of mitochondrial and metabolic vestiges of conflict mediation. Many of these vestiges should lead back to the mechanisms of chemiosmosis, which are suggested to be central to evolutionary conflict in previous chapters. This may provide a framework to rationalize at least some seemingly baroque cellular and molecular natural histories.

Nonfunctional vestiges, however, are unlikely to be found. Rather, any vestiges extant today would likely have been co-opted into new functions for two reasons. First, if they were not, they would not survive mutational decay. Second, evolution is much more likely to co-opt an old mechanism into a new function than to actually invent something new [2]. This is an aspect of modern biology that has received little attention, but potential examples may abound, as will be suggested in this chapter.

Impetus for this viewpoint came with the discovery that mitochondria were intimately involved in programmed death of eukaryotic cells [3, 4]. Overnight, it became something of a cottage industry to rationalize this seemingly counterintuitive result in evolutionary terms [5–7]. This example remains instructive [8], and it is worth briefly recounting here. In a number of animals and perhaps other eukaryotes, release of cytochrome c is a key step in the initiation of programmed cell death via the mitochondrial pathway [3, 4, 9]. Cytochrome c is a well-known protein

N. W. Blackstone, *Energy and Evolutionary Conflict*,
https://doi.org/10.1007/978-3-031-06059-5_11

component of the mitochondrial electron transport chain (Chap. 2), and the discovery of its role in programmed cell death was greeted with amazement. Why would an electron carrier of the mitochondrial electron transport chain signal cell death? The answer likely involves both redox signaling and levels-of-selection conflicts that occurred in the eukaryotic stem group [10, 11]. Descendants of proteobacteria, protomitochondria no doubt were well equipped with redox signaling mechanisms. Sophisticated mechanisms such as the two-component signal-transduction mechanism described in Chap. 2 cannot be expected to function in this "new" chimeric organism because the host was a very different kind of bacterium. "Generic" signaling mechanisms [12] would have a greater likelihood of functioning under these circumstances.

Consider again the early days of complex cells (Chap. 7). If the environment became stressful for the host and thus unfavorable for replication, yet plenty of substrate remained available, metabolic demand of the host would diminish, and the ATP/ADP ratio of the protomitochondria would approach one. Oxidation of substrate, however, would continue until the electrochemical gradient of the protomitochondrial inner membranes was maximal and the electron carriers were highly reduced. In such circumstances, these electron carriers would freely donate electrons to diatomic oxygen. In sufficient quantities, reactive oxygen species (ROS) thus formed can trigger genetic mutations in the host, leading ultimately to sexual recombination [13]. As a result, the protomitochondria could thus "engineer" genetically novel host cells that were products of sexual recombination and potentially better able to cope with the original stress. Subsequent to selection, rapid growth and high metabolic demand of the successful host and symbiont population could once again ensue. The electron carriers of the protomitochondrial inner membranes would then become relatively oxidized and only low levels of ROS would be produced.

To enhance this redox signal, the symbiont could manipulate ROS formation by completely blocking the electron transport chain. Cytochrome c release should be viewed in this context, since its release from the electron transport chain effectively blocks electron flow. Cytochrome c release would have the same effect as low metabolic demand of the host—electron carriers "upstream" of cytochrome c would become highly reduced and ROS formation would become maximal. Note that the two major sites of ROS formation in the electron transport chain (complexes I and III) are both upstream of cytochrome c (Chap. 2).

As the relationship between the host and the mitochondria became better established, conflicts between the units of selection would require mediation [14]. To accomplish this, there would be selection for more refined signaling mechanisms. "Generic" signaling mechanisms can be effective, but the lack of precision can have fitness costs. In particular, signaling with high levels of ROS will have a certain amount of collateral damage as otherwise valuable molecules are damaged and require repair or replacement. In this case, selection should favor making cytochrome c itself the signal, rather than the potentially dangerous ROS. ROS formation could thus be diminished without impairing the transduction of the signal (e.g., in addition to deploying antioxidants and uncoupler proteins, by releasing cytochrome c molecules from the mitochondrial intermembrane space—such molecules are not actively participating in electron transfer and thus their release does not block electron flow and upregulate ROS).

This mechanism for triggering sex in a unicellular organism may have been co-opted into a mechanism for cell death after multicellularity evolved. In particular, the dynamics of adding a third level of selection (groups of eukaryotic cells) could favor such co-option [13]. In such a group, selection on the higher-level unit (i.e., the group of cells) may favor eliminating, rather than repairing, damaged cells. Linking a pathway that once led to a repair mechanism (i.e., sex) in a single cell to a pathway leading to elimination of damaged cells (i.e., cell death) in a group would seem to be a rather straightforward co-option.

The history of eukaryotic cells may be littered with similar vestiges of redox signaling. Initially, these mechanisms were used by symbionts to exploit and manipulate host cells. Subsequently, these mechanisms evolved to mediate levels-of-selection conflicts and stabilize the symbiosis on which the eukaryotic cell was built. Finally, these mechanisms are co-opted by groups of eukaryotic cells to suppress defecting cells. In this lengthy evolutionary process, genes initially involved in these mechanisms may subsequently have been suppressed, while at the same time, novel genes may have been recruited. While mechanisms of programmed cell death have been examined in this context, there may be many more such examples.

Indeed, such a framework may be of general utility when considering the evolution of mechanisms of molecular cell biology. Consider the *p53* family of tumor-suppressor genes [15, 16]. This gene family mediates adaptive response to genotoxic stress and is usually compromised in human cancers. In a recent article [16], the authors plausibly argue that these tumor-suppressor functions are likely derived from more primordial functions related to meiosis. However, their rationale for this—that cancers are rare during the normal lifespan of most animals—is unconvincing. Cancers might be rare because they are strongly selected against, and animal lifespan evolved in the context of such selective forces [17]. Further, there is a much better rationale for the same conclusion—cancer is a disease of multicellular, not unicellular, organisms. Increasingly, at least some of the proteins involved in complex signaling pathways in multicellular eukaryotes are found in unicellular eukaryotes performing other tasks [18]. Since unicellularity clearly preceded multicellularity in the history of life, there can be a strong a priori expectation that any unicellular function also preceded a necessarily multicellular one.

In the context of the foregoing discussion of the evolution of cell death, the linking of *p53* function to meiosis may be suggestive: unicellular signaling with *p53* or something like it may have evolved in the context of host-symbiont interactions. Certainly, the role of modern *p53* involves considerable metabolic regulation. In its role as a modulator of gene expression, *p53* can affect the rate of both glycolysis and oxidative phosphorylation [19]. Additionally, *p53* can directly activate the expression of genes coding for antioxidant proteins [15, 19–21]. Other studies implicate the mitochondrial electron transport chain in the *p53* response [22–23]. All of these actions of *p53* suggest an ancestral unicellular function as compared to the derived tumor-suppressor function. Further study might falsify this hypothesis, but nevertheless this remains a reasonable a priori approach.

The search for metabolic vestiges of conflict mediation in the molecular biology of modern organisms may be rewarded by other exemplars, as suggested by the following eclectic selection.

The Curious Case of STAT3 For a multicellular organism to function, cells must communicate with each other. Cytokines are a large and diverse family of molecules used in such between-cell signaling in animals. These molecules and their associated receptors and signaling pathways have been particularly well studied in mammals in which they are implicated in diseases such as cancer [24].

Signal transducer and activator of transcription (STAT) proteins are typically latent in the cytoplasm until activation by extracellular signaling molecules such as cytokines [25]. The signaling proteins that activate STATs also include growth factors and even some simple peptides. These signaling proteins bind to cell-surface receptors and activate tyrosine kinases, which subsequently phosphorylate STAT proteins. STAT proteins that are so activated then accumulate in the nucleus and initiate transcription, ultimately affecting the phenotype of the cell. Particularly well studied are the Janus kinases (JAKs) and their STAT targets. The canonical JAK-STAT pathway is an important example of a complex signaling pathway with broad relevance to human health and disease [26].

One of the members of the STAT family, STAT3, was first described for its DNA-binding activity in IL-6 cytokine-stimulated hepatocytes. The protein was found to be structurally similar to other STATs. In response to cytokine stimulation, a Janus kinase mediates tyrosine phosphorylation, which occurs at a single site close to the carboxy terminus [27]. Activated STAT3s dimerize, translocate to the nucleus, and initiate DNA binding. However, STAT3 can also be activated by serine phosphorylation at a site in the transactivation domain. The role of serine phosphorylation in transcriptional activity has remained rather ambiguous [27]. In comparison to other STATs, the function of STAT3 also seems unique, with data suggesting a general role in regulating cellular homeostasis [26, 28]. At the same time, the diverse roles of STAT3 raised questions of how a single transcription factor could be involved in such seemingly contradictory responses [27].

Recent work [29–31] has clarified the ambiguity surrounding the function of STAT3 by showing that it actually has two distinct functions. Early clues were provided by the interaction of STAT3 with GRIM-19, which is a component of complex I of the mitochondrial electron transport chain (Chap. 2). Wegrzyn et al. [31] carry these observations several steps further and show that some of the STAT3 in a cell indeed localizes to mitochondria. Additional evidence indicates that STAT3 is a component of complex I and possibly complex II of the mitochondrial electron transport chain. Using cells deficient in STAT3, Wegrzyn et al. [31] show that the capacity for oxidative phosphorylation (which is carried out by the mitochondrial electron transport chain) is diminished in these cells as well. A functional role of STAT3 in complex I and II is suggested. Indeed, serine phosphorylation seems to be integral to this mitochondrial function.

Gough et al. [29] follow up on this work. Augmented STAT3 activity is associated with numerous human tumors, yet such an observation seems inconsistent with STAT3 acting solely as a transcription factor. Using Ras-dependent oncogenic transformation as an exemplar, the authors show that this transformation was dependent on STAT3. While tyrosine phosphorylation is not required, serine phosphorylation of STAT3 is critical for this transformation. Further data support the hypothesis that Ras transformation requires non-transcriptional and nonnuclear STAT3. This suggests a mitochondrial role for STAT3, raising the possibility of a connection between STAT3 activity and the abnormal mitochondrial metabolism that characterizes cancer cells (see Chap. 10). Indeed, mitochondrial STAT3 appears to contribute to Ras-dependent cellular transformation by altering the activity of complexes of the electron transport chain as well as somehow upregulating glycolysis.

All of this groundbreaking work on STAT3 was apparently carried out and reported without any reference to the evolutionary history of eukaryotic cells. One might surmise from reading this literature that such an evolutionary view could not possibly add any insight to the still ongoing investigation of the curious case of STAT3. Such a judgment would be premature without at least a consideration of this evolutionary history and what perspective it could add to the STAT3 story [32].

Considerations of the evolutionary context for the function of STAT3 begin with the now widely accepted mitochondrial endosymbiosis. As described in Chap. 7, a number of hypotheses have been proposed regarding the nature of the initial association and the capabilities of the original host and symbiont [33, 34]. Some areas of broad agreement nevertheless have emerged. The mitochondrial symbiosis is generally viewed as a seminal event in the origin of complex cells [1, 35, 36]. This symbiosis created the principal compartment for eukaryotic metabolism, but as always, the lunch is never free. The early stages of this symbiosis were likely very different from the relative harmony seen in modern eukaryotic cells. Because mitochondria were evolutionary units capable of heritable variation, levels-of-selection synergies and antagonisms no doubt ruled the emerging features of the eukaryotic cell. Much of the cooperation and conflict that occurred related to a functional difference between the symbiont and the host: at or soon after the onset of the symbiosis, the symbiont possessed a functional electron transport chain, while the host lacked this feature [37]. When analyzing any feature of modern eukaryotes, this evolutionary history should be kept in mind. Mitochondrial signaling pathways may remain as vestiges of ancient levels-of-selection conflicts [13].

Now consider the particular case of STAT3. Two evolutionary interpretations are possible, and each will be discussed. First, STAT3 may have originally been a mitochondrial protein that was co-opted into a new function of manipulating the host to the advantage of the symbiont. Once mitochondria became obligate symbionts and thus part of a new higher-level evolutionary unit (the eukaryotic cell), they would no longer directly interact with the environment. Selection would favor symbionts that could trigger particular host responses to stimuli if those responses subsequently increased symbiont fitness. (Think of small children riding in the back seat of a car during a long trip: a well-timed "Daddy, I'm hungry!" leads to Daddy stopping the car and feeding them.) In this context, a mitochondrial protein that could act as a transcription factor for host DNA would be an invaluable tool. When symbionts detected certain metabolic signals that were ultimately environmental, this transcription factor could be activated, perhaps by phosphorylation, and could move to the nucleus where appropriate gene activity would be initiated. In this case, "appropriate gene activity" would benefit the symbionts; it might also benefit the host under a certain range of conditions or have no effect on the host at all. If this gene activity was detrimental to the host, i.e., if there is a conflict between what selection favors at the level of the symbiont and at the level of the host, then the evolutionary calculus would become more complex.

Electron transport chains are typically the locus of not just energy conversion, but environmental sensing as well. Bacteria illustrate this point particularly well, as described in Chap. 2. Here it is worth recounting the basic principles of two-component signal-transduction systems as illustrated by the Arc system of *Escherichia coli*. Two proteins, ArcA and ArcB, are involved. ArcB is a transmembrane sensor kinase with a loop exposed to the cytoplasm. This cytoplasmic loop contains a conserved histidine residue that can be autophosporylated in response to metabolic conditions. This phosphorylation occurs in response to the oxidation state of quinone electron carriers (i.e., if they are saturated with elections or not). Quinones are part of the electron transport chain (Chap. 2). Oxidized forms of quinone inhibit autophosphorylation [38]. On the other hand, if quinones are reduced, this inhibition is removed. The oxidation state of quinones is sensitive to environmental conditions. In the presence of substrate and ADP, quinones remain relatively oxidized, and autophosphorylation is inhibited as long as electron transport to the terminal electron acceptor (oxygen) is possible. If oxygen is not available, electrons "back up" on the electron carriers of the electron transport chain, and these carriers become reduced. Autophosphorylation then ensues. Subsequent to autophosphorylation, ArcB transphosphorylates the second component, ArcA, which is a global regulator of transcription. When phosphorylated, ArcA represses the expression of many genes whose products are involved in aerobic respiration and activates many of the genes whose products are involved in anaerobic fermentation. In this way, the bacterium adapts its metabolism to the environmental conditions.

Mitochondria are descended from bacteria not unlike *E. coli*. Primitively, they are expected to have employed similar environmental sensing mechanisms. Consider STAT3 in this context. Evidence suggests that if it is activated by serine phosphorylation, it is a component of mitochondrial complexes I and II. On the other hand, if it is activated by tyrosine phosphorylation, it is a nuclear transcription factor. Such a juxtaposition of functional roles suggests that it may have originally been an environmental sensor for the mitochondrial electron transport chain. Under the appropriate metabolic conditions, it could quickly be converted into a nuclear transcription factor, modifying gene activity to suit the metabolic circumstances of the mitochondria. While initially there may have been exploitative aspects to this interaction (i.e., favored by selection on mitochondria, but possibly disadvantageous to the host), simultaneously there would have been strong selection on a host to maintain mitochondria in good functional condition. Put another way, a host responsive to signals that facilitate mitochondrial metabolism would reap an energy dividend in the form of ATP generation. This energy dividend could then be used in faster host replication and higher host fitness. Thus, the system of a STAT3 mitochondrial sensor/nuclear transcription factor could have quickly evolved into a mutually beneficial signaling pathway.

The second evolutionary interpretation of STAT3 signaling begins with the alternative hypothesis that it was originally derived from the host. Given that the principle functional difference between the host and the symbiont was the presence of the electron transport chain, products and by-products of this chain could be used by the symbiont population within a single host to manipulate their host [37]. For instance, reactive oxygen species, a by-product of respiration, could be used to trigger recombination and whole cell fusion in the host, thus providing the symbionts with new habitat [13]. Ultimately, such manipulation would destabilize the symbiosis because new and "selfish" variant symbionts would continuously evolve, sacrificing the group-level benefits of cooperation for short-term gains in individual-level fitness. For a stable symbiosis to emerge, the higher-level unit (which includes the host as well as the entire population of symbionts) must evolve mechanisms to hold such selfish variant mitochondria in check [14]. In modern mitochondria, these mechanisms are many and various, perhaps most notably shifting the bulk of the mitochondrial genome to the nucleus. Moving these genomes to the nucleus diminishes the amount of heritable variation available to produce selfish lower-level units. The mitochondrial electron transport chain in particular has seen most of the genes coding for components of the complexes I–V moved to the nucleus. With some important exceptions [39, 40], regulation of respiration is too critical to be left to the control of individual symbionts. Since it required the evolution of a set of transporters to import gene products from the nucleus, gene loss from protomitochondria may nevertheless have taken a relatively long period of time to accomplish. Early in the symbiosis, inserting a host protein into the electron transport chain to allow community-level regulation of respiration may have been critical and strongly selected for. STAT3 is a plausible candidate for such a regulatory protein. Ultimately, eukaryotic cells with this regulatory protein established a stable symbiosis, while those without it succumbed to the selfish manipulation of mitochondrial variants.

Since the canonical JAK/STAT signaling pathway is involved in signaling between cells of the same multicellular organism, this pathway is expected to have evolved later in the history of eukaryotes, perhaps as animals themselves evolved. Again, at least some of the proteins involved in complex signaling pathways in multicellular eukaryotes are found in unicellular eukaryotes performing other tasks [18, 41]. In this context, host-symbiont signaling may provide a plausible functional origin of the STAT family of proteins. While the function of modern STAT3 is reminiscent of this original function, it and all of the other STAT proteins likely were extensively modified subsequent to this origin. It would be misleading to consider STAT3 an "ancestor" to other STAT proteins, just as it is misleading to consider any extant organism an "ancestor" of any other extant organism.

Indeed, the noncanonical (i.e., serine phosphorylation) pathway of STAT3 clearly participates in critical functions of the immune system, which of course are characteristic of multicellular organisms. For instance, the activation of macrophages by microbial components depends on this pathway [42]. Once these components are detected, macrophages take on many of the metabolic characteristics of cancer cells (Chap. 10), becoming more glycolytic, increasing reactive oxygen production, and directing Krebs cycle intermediates into biosynthetic pathways. This metabolic reprogramming is mediated by serine phosphorylation of STAT3, much like the Ras-dependent cellular transformation described above.

Mitochondria Make Waves Ca^{2+} is a highly versatile signal in eukaryotic cells that regulates many different functions [43, 44]. The key to this signaling is that the intracellular concentration of calcium is maintained at a much lower level than that of the extracellular environment. For instance, in mammals, the concentration within a cell approaches 100 nM, as compared to the extracellular concentration of 1 mM [45]. Various pumps, channels, exchangers, and binding proteins maintain the low intracellular concentration. Continuous high levels of calcium have a detrimental effect, e.g., leading to the precipitation of calcium phosphate. Against this low intracellular background, influx from the extracellular environment or intracellular stores results in pulses or waves of Ca^{2+}. These pulses can be tuned in an amazing number of ways, rendering the calcium ion a key "second messenger" in multicellular animals. As pointed out by Jacobson and Duchen [43]:

> Cellular calcium signaling seems to underpin an almost indecent array of processes that involve the transition of cell activity from quiescent to active—be it contractions of muscle—smooth, cardiac, skeletal, secretion—of neurotransmitters, of hormones, or immunoactive compounds and cytokines, control of the cell cycle, of motility and so on and on. It has become standard practice in recent discussions on mitochondrial involvement in cellular calcium handling to talk of a 'renaissance' in understanding of the roles of this remarkable organelle in relation to cellular calcium signaling.

While the many permutations of calcium signaling will not be reviewed here, describing at least one example will serve as a useful illustration of the general process [44, 45]. The main route of calcium release from intracellular stores involves the inositol trisphosphate receptor (IP_3R), a transmembrane protein located on the endoplasmic reticulum and Golgi membrane. When extracellular soluble agonists bind a G-coupled protein receptor on the plasma membrane, phospholipase C (PLC) isoforms are activated. Hydrolysis of a precursor by PLC produces IP_3, which then binds to IP_3R and induces its opening and the release of Ca^{2+}. Calcium waves then ensue. IP_3R is not the only channel that can release calcium from the ER or other intracellular stores, but its function will serve as a general example in the following discussion.

In the IP_3 pathway and in other calcium signaling pathways, mitochondria have been perceived as everything from irrelevant to central [46]. As described in the fascinating historical account by Carafoli [46], experiments in the 1950s suggested that calcium could stimulate respiration in isolated mitochondria, much like ADP [47]. In post-chemiosmotic hindsight, these early data now make considerably more sense (see Chap. 2): an influx of positively charged calcium ions diminishes the transmembrane gradient, and this stimulates oxidative phosphorylation to recharge the gradient. Subsequently, a considerable amount of research focused on the ability of mitochondria to take up large quantities of calcium ions, particularly under pathophysiological conditions. Connections between calcium and bioenergetics re-emerged from data showing that several Krebs cycle enzymes could be activated by Ca^{2+}. From the perspective of cell- and organism-level calcium signaling, however, the data that were decisive at the time were published in the 1970s. These data showed that the affinity of the mitochondrial uniporter for calcium was quite weak. At the concentrations prevailing in cells (100 nM), the mitochondrial uniporter would be only marginally active. In contrast, the endoplasmic reticulum was found to have a much higher affinity for calcium and was thus thought to be much more central to cellular calcium regulation. With the discovery of IP_3 signaling, mitochondria were generally deemed to be irrelevant to cellular calcium signaling.

While some discordant data continued to indicate a role for mitochondria in cellular calcium signaling [43, 46], the prevailing view remained fixed for a number of years. By the 1990s, however, the discordant data could no longer be ignored [48]. A role for mitochondria in cellular calcium signaling was reconciled with their low affinity for calcium by recognizing that "Ca^{2+} microdomains" could exist within the cell. Thus, when IP_3 binds to IP_3R, the massive pulse of calcium being released by the endoplasmic reticulum could easily activate mitochondria that are nearby. A considerable role for calcium in host-symbiont signaling becomes entirely plausible [49]. Further, the data showed that not only do mitochondria congregate in the appropriate areas, but they also form a specialized junction with the endoplasmic reticulum: the mitochondria-associated ER membrane (MAM) [50, 51].

The structure of MAM is complex with key proteins arrayed across three layers of membranes. In addition to IP_3R, sarco-ER Ca^{2+} ATPase (SERCA) are found on the ER membrane. SERCA is an ATP-driven pump which refills the internal calcium stores of the ER after a pulse exits IP_3R. On the mitochondrial outer membrane, voltage-dependent anion channels (VDAC) allow calcium and ADP to enter the intermembrane space and ATP to exit. On the mitochondrial inner membrane, adenine nucleotide translocator (ANT; sometimes referred to as ADP/ATP carriers) imports ADP and exports ATP, and the mitochondrial calcium uniporter (MCU) imports Ca^{2+} and exports protons, while Na^+/Ca^{2+} exchangers (NCX) import sodium and export calcium. MCU has more recently been characterized [52, 53]. The nexus of membrane proteins that is MAM thus allows calcium to exit the ER via IP_3R, pass through the mitochondrial outer membrane via VDAC, exit the mitochondria via NCX and VDAC, and reenter the ER via SERCA. Meanwhile, the ADP that is formed from ATP to power SERCA can pass through VDAC and be exchanged for ATP at ANT. The centrality of mitochondria to cellular calcium signaling could not be more explicit.

The physical and functional coupling of MAM allows the mitochondrial metabolic state to be telegraphed in calcium waves. For instance, in mitochondria that are provided with sufficient substrate, entry of moderate amounts of calcium will lower the membrane potential and cause an immediate activation of oxidation of substrate to rebuild this potential. Several of mitochondrial enzymes used in processing substrate are activated by calcium. These include pyruvate dehydrogenase, which is discussed below because of its sensitivity to insulin. These activated enzymes can then maintain the flow of electrons to the electron transport chain, which can continue to build membrane potential. Using this potential, ATP can continue to energize the nascent calcium wave. On the other hand, consider such calcium signaling when mitochondria are deprived of substrate. Such mitochondria would be unable to rebuild their membrane potential or convert ADP to ATP. The cellular calcium signal would be stillborn.

These general considerations suggest that while many aspects of calcium signaling in modern cells may be more recently derived, the process itself has roots in the formation of the eukaryotic cell. While complex associations with the ER and even the ER itself likely did not exist, mitochondria could have similarly energized calcium waves emanating from the plasma membrane [49]. In the presence of substrate, a calcium signal could be energized by mitochondria, while at the same time, ATP would increase and ROS would decrease. If mitochondria were deprived of substrate, however, the calcium signal would not be energized. Hungry mitochondria could not be ignored. Calcium, along with ROS and ATP, may have been one of the levers that mitochondria used to manipulate their hosts. The many facets of calcium signaling in modern cells may all be vestiges of this ancient host-symbiont interaction.

VEGF-B: A Link Between Angiogenesis and Mitochondrial Metabolism Human cancers can be viewed as the selfish replication of an individual cell (Chaps. 9 and 10). As the selfish cell produces a clone of daughter cells, the growing tumor needs to gain access to the organism's food supply. Typically, this is done by triggering the growth of vascular tissue [54]. Indeed, all human oncogenes and tumor-suppressor gene pathways have been implicated in angiogenesis, either directly or indirectly [55].

Vascular endothelial growth factor (VEGF) proteins are major regulators of angiogenesis as well as other aspects of endothelial cell physiology in mammals [56]. This protein family includes VEGF-A, VEGF-B, placental growth factor (PlGF), VEGF-C, and VEGF-D. The growth factors bind their cognate receptors VEGFR-1, VEGFR-2, and VEGFR-3, which are found on the cells of the vascular endothelium. In some of the literature describing this important protein family, VEGFR-1, VEGFR-2, and VEGFR-3 may be referred to as FLT1, FLK1, and FLT4, respectively [57, 58]. In any event, VEGF-A binds to VEGFR-1 and VEGFR-2 and other receptors and induces proliferation, sprouting, migration, and tube formation [56]. VEGF-A is also a potent survival factor for endothelial cells and triggers vasodilation through the induction of nitric oxide synthase. PlGF is expressed predominantly in the placenta, heart, and lungs [56]. VEGF-C and VEGF-D induce mitogenesis, migration, and survival of endothelial cells [56].

The precise role of VEGF-B has remained elusive [56]. VEGF-B seems to have little effect on angiogenesis except in the heart [57]. Hagberg et al. [59] reveal a connection between the VEGF family of proteins and mitochondrial metabolism. Subsequent to the appropriate metabolic signal, VEGF-B and a battery of mitochondrial genes are upregulated. The mitochondrial proteins thus produced are involved in the β-oxidation of fatty acids. In this process, fatty acids are oxidized to acetyl coenzyme A (CoA). Acetyl-CoA feeds into the Krebs cycle and ultimately builds the mitochondrial inner membrane proton gradient, triggering the formation of ATP (Chap. 2). However, if free fatty acids are to be made available to mitochondria in cells of the tissue, they have to be transported from the blood across the vascular endothelium. These data suggest that VEGF-B regulates the transcription of genes for these fatty acid transport proteins. The expression of *Vegfb*, in turn, is regulated by peroxisome proliferator-activated receptor γ coactivator 1α (PGC-1α), a major regulator of mitochondrial energy metabolism, via estrogen-related receptor α (ERRα) [60].

VEGF-B may thus be an important regulator of mitochondrial metabolism. Plausibly, its original function was to activate fatty acid transporter proteins in a unicellular organism. Scenarios for the evolution of the VEGF family of proteins should take this into account. VEGF-B may have originated as a symbiont protein that connected mitochondria to an external food supply. Alternatively, it may have originated as a host protein that regulated the availability of food for mitochondria. If unicellularity is considerably more ancient that multicellularity, a compelling case can be made that the mitochondrion-related function represents the ancestral state for the entire VEGF family of proteins.

As with STAT3, further tests of this hypothesis should keep in mind that VEGF-B may be no more similar to this ancestral protein than any of the VEGF family proteins. The function of VEGF-B is suggestive of the ancestral, unicellular function. Nevertheless, all of the members of VEGF family of proteins have evolved an equal amount of time from this putative ancestor. None of the proteins existing today can be thought of as an "ancestor," just as no organism extant today can be thought of in this way.

In any event, the parallels to STAT3 are striking. Both STAT3 and VEGF-B are members of protein families with important roles in human health and disease. Nevertheless, the exact functions of both of these individual proteins remained obscure as compared to better studied family members. As these functions were elucidated, it became clear that both STAT3 and VEGF-B were important regulators of mitochondrial metabolism. Because the evolution of complex cells preceded the evolution of animal multicellularity, in both cases, these mitochondria-related functions may constitute the ancestral functions for both protein families.

This is an intriguing point. In cases as diverse as programmed cell death and STAT and VEGF proteins, basic regulatory mechanisms may date to the origin of the eukaryotic cell. Metabolic regulation in particular evolved from mediating levels-of-selection conflicts. When eukaryotes become multicellular, these within-cell pathways may then have been co-opted into between-cell pathways. The success of eukaryotes in mediating within-cell conflicts may have preordained their success in mediating between-cell conflicts and thus achieving multicellularity. Certainly, eukaryotes have achieved notable success as multicellular organisms [61]. The challenges that had to be overcome in forming a higher-level unit out of a community of energy-converting lower-level units may have given eukaryotes a remarkable toolkit to overcome such conflicts in subsequent evolutionary transitions. Perhaps this is the evolutionary version of the old adage—what doesn't kill you makes you stronger, or at least more evolvable.

Insulin and the Power of Substrate As discussed in Chap. 2, excessive reactive oxygen species (ROS) are central to the pathological effects of diabetes and related human diseases [62]. An imbalance in ROS production, in turn, is caused by exposure of mitochondria to too much substrate relative to the metabolic demand of their host cell. This shifts the redox state of the mitochondrial electron transport chains in the direction of reduction and amplifies ROS formation. The breakdown in the regulation of glucose levels can thus have severe consequences. In some cases, this breakdown is due the hormone insulin, which is central to regulating carbohydrate and fat metabolism in mammals. In healthy human individuals, insulin is produced by islets of Langerhans cells in the pancreas. Insulin is released into the blood and binds to receptors on target cells. In these cells, GLUT4 glucose transporters then translocate from the endoplasmic reticulum to the plasma membrane where they catalyze the movement of glucose across the membrane and into the cell [63]. Among other effects, insulin also activates the mitochondrial pyruvate dehydrogenase complex, which contributes to transforming pyruvate into acetyl-CoA [64]. Pyruvate is of course the product of glycolysis, and acetyl-CoA is the substrate for the Krebs cycle (Fig. 11.1). Thus, activation of the pyruvate dehydrogenase complex is a pivotal step in mitochondrial metabolism. In parallel with previous examples, proteins involved in between-cell signaling have within-cell—and in particular mitochondria-related—functions. Again, since unicellular organisms preceded multicellular ones, arguably the within-cell functions are the primitive ones. Subsequent to the evolution of multicellularity, the within-cell functions of insulin or insulin-like proteins were plausibly co-opted into new between-cell functions.

The lessons of insulin, however, are broader. Within a cell, glucose metabolism is a complex multistep pathway (Fig. 11.1). Altering the supply of substrate at one step has cascading effects throughout the entire pathway, and indeed, throughout the entire cell. A series of studies by Veech and colleagues emphasizes this point [65–67]. These studies compare the effects of insulin to ketone bodies. The latter comprise acetone, acetoacetate, and β-hydroxybutyrate. Ketone bodies are well known because of the so-called ketogenic diet, which is discussed in Chap. 10. Subject to this high-fat, low-carbohydrate diet, the mitochondria in the human liver burn fatty acids by β-oxidation. An excess of acetyl-CoA is generated and converted into ketones, which circulate in the blood. Ketone bodies are then utilized as an energy source, particularly by the brain, since fatty acids do not cross the blood-brain barrier [68]. The ketogenic diet has been used for nearly a century to treat pediatric epilepsy [68–70]. Mild ketosis might also be effective for neurodegenerative diseases [71]. As discussed in the previous chapter, the ketogenic diet may potentially limit some cancers, at least in part because of its effects on dietary glucose.

In comparison to these chronic effects, the acute effects of ketone bodies are interesting as well. With a rat heart model, β-hydroxybutyrate and acetoacetate were applied in a physiological ratio (4/1) [65–67]. Metabolism of ketones led to a shift in the $NAD^+/NADH$ ratio in the direction of reduction, an increase in the

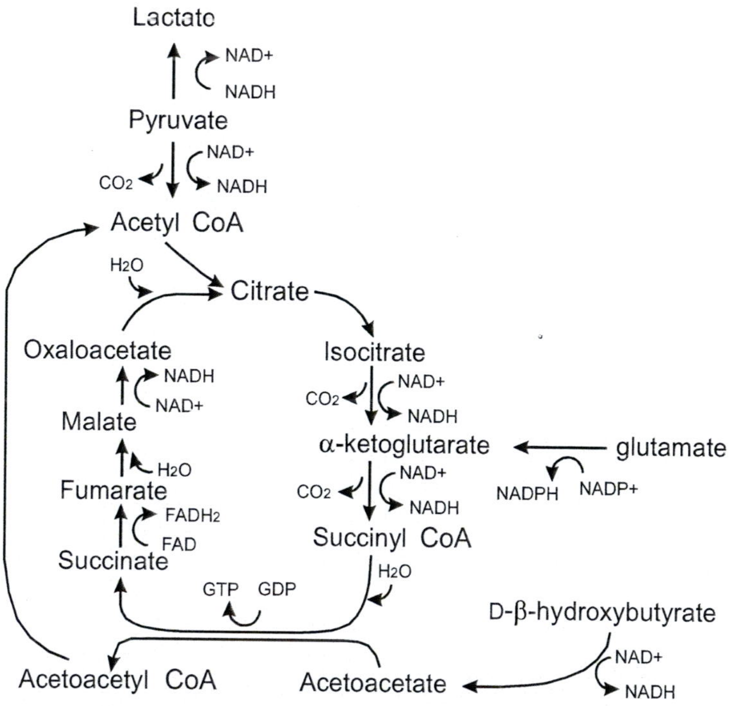

Fig. 11.1 Simplified diagram of the Krebs cycle and related metabolism. Pyruvate produced from glucose can be converted to lactate or broken down in mitochondria. The first step in this later process is catalyzed by pyruvate dehydrogenase and results in acetyl-CoA, which is fed into the Krebs cycle by reacting with oxaloacetate. In each turn of the cycle, the products include two molecules of CO_2, three molecules of NADH, one molecule of GTP, and one molecule of $FADH_2$. NADH and FADH2 are oxidized by the electron transport chain to build the transmembrane electrochemical gradient. (Modified from Sato et al. [66])

metabolites of the first third of the Krebs cycle, and a shift in CoQ pool in the direction of oxidation [65–67]. These metabolic changes largely parallel the acute effects of insulin, despite what must be different modes of action. For instance, insulin vastly increases intracellular glucose concentrations, but addition of ketones, an alternative substrate, lessens the use of glucose by the cell. A commonality may be the effects on acetyl-CoA, the substrate fed into the Krebs cycle (Fig. 11.1). Insulin, by increasing the concentration of glucose and activating the pyruvate dehydrogenase complex, also unsurprisingly increases the supply of acetyl-CoA. In the case of ketone bodies, β-hydroxybutyrate is converted to acetoacetate (accompanied by the reduction of NAD^+ to NADH). Acetoacetate is then converted to acetyl-CoA (Fig. 11.1). Insulin and ketone bodies thus achieve the same effects on acetyl-CoA via very different enzymatic pathways. As summarized by Sato et al. [66]:

> A physiological ratio of ketone bodies at a total concentration of only 5 mM thus was able to duplicate most of the acute effects of insulin by providing increased amounts of acetyl

> CoA by a mechanism not involving the activation of pyruvate dehydrogenase multienzyme complex…, and at the same time induce similar changes in the contents of TCA cycle metabolites and mitochondrial redox couples. This suggests that the major acute effect of insulin results simply by increasing the supply of mitochondrial acetyl CoA....

These results raise the possibility that the moderate ketosis associated with some forms of diabetes may arise as physiological compensation for the lack of insulin [66]. Thus, while insulin illustrates another example in which between-cell metabolic signaling may have been co-opted from within-cell signaling, there is also a broader message here concerning the power of substrate. Insulin-like proteins may accomplish their signaling at least in part by tipping the balance of substrate in a particular direction. Metabolic regulation and signaling pathways thus intertwine in an endless knot.

Insulin-like signaling pathways seem to be widespread in animals and have attracted considerable attention because of connections between aging, metabolism, and development [72]. Parallels between the effects of diminished insulin/insulin-like growth factor-I signaling and calorie restriction in diverse animal taxa further support the connections between substrate and signaling suggested above.

Innate Immunity and Mitochondria Multicellular organisms constantly survey tissues for foreign particles and then seek to eliminate them (e.g., the innate immune response in animals [73] and the hypersensitive response in plants [74, 75]). Reactive oxygen species (ROS) are typically used in these interactions to degrade foreign particles, to initiate cell death, or for other functions [74–76]. In animals, generation of these ROS was thought to involve entirely non-mitochondrial mechanisms (e.g., NADPH oxidases) [76]. In contrast to this earlier view, recent reports now implicate widespread involvement of mitochondrial ROS in this response [77, 78]. As suggested by Zhou et al. [79]:

> There is a large amount of literature proposing a link between mitochondrial malfunction, ROS and chronic inflammatory diseases. Damage to mitochondria is now understood to have a role in the pathogenesis of a wide range of seemingly unrelated disorders.

Sonoda et al. [77] provide an example of innate immune function that ties together several of the signaling pathways discussed above. Interferon-γ (IFN-γ), a pro-inflammatory cytokine, triggers the antibacterial activities of macrophages through the activation of the JAK/STAT1 pathway (see above). Among other effects, IFN-γ induces the expression of estrogen-related receptor α (ERRα), a ligand-activated transcription factor. ERRα is a target of peroxisome proliferator-activated receptor γ coactivator-1α (PGC-1α) and PGC-1β, the master regulators of mitochondrial metabolism. Sonoda et al. [77] suggest that PGC-1β and ERRα act together as a molecular switch to promote mitochondrial function during macrophage activation. During times of inactivity, this switch is in the "off" position, so unwanted mitochondrial ROS are limited. Uncoupling protein 2 may also be used to limit ROS production in inactive cells. The switch is turned "on" by IFN-γ activation of the JAK/STAT1 pathway and the direct binding of STAT1 to conserved elements in the PGC-1β promoter. Upregulation of PGC-1β activates ERRα, which affects broad aspects of mitochondrial biology, including fatty acid oxidation, mitochondrial biogenesis, oxidative phosphorylation, and so on. Copious ROS are produced as a by-product of this metabolic activity.

In this context, it is important to qualify the connections between metabolic state and ROS, which was briefly mentioned in Chap. 10. Greater metabolic demand will act as a "starter's pistol" for oxidative phosphorylation. Greater oxidation of substrate and reduction of the terminal electron acceptor will typically leave the electron carriers relatively oxidized. All other factors being equal, formation of ROS under these circumstances will be moderate. If metabolic demand ceases, electrons will "back up" on the electron transport chain, and ROS formation will increase. In some ways, these generalities would seem to contradict the data from the previous paragraph. This contradiction can be resolved by recognizing that in the previous paragraph, all factors did not remain equal. Production of more electron transport chains per mitochondrion and more mitochondria per cell will inevitably increase the production of ROS. Regardless of their metabolic state, mitochondrion-rich cells will exhibit considerable amounts of ROS [78].

Other signaling pathways that activate macrophages similarly upregulate mitochondrial ROS. West et al. [80] describe the engagement of Toll-like receptors (TLR1, TLR2, and TLR4) in this context. When bone marrow-derived macrophages were stimulated with agonists to these Toll-like receptors, mitochondrial ROS increased. Indeed, mitochondria were recruited to phagosomes containing foreign particles. This latter response involved translocation of a TLR-signaling adaptor to mitochondria. There, the signaling adaptor (TRAF6, tumor necrosis factor receptor-associated factor 6) engaged another protein, ECSIT (evolutionarily conserved signaling intermediate in Toll pathways). ECSIT has a role in the assembly of complex I of the mitochondrial electron transport chain [81]. Possibly, ECSIT modulates the amount of ROS emanating from complex I. Once again, signaling related to a critical function in multicellular organisms has led back to the mitochondrial electron transport chain.

Still another constituent of the innate immune system, the inflammasome, also exhibits a mitochondrial connection. The NLRP3 (nod-like receptor P3) inflammasome is a multiprotein complex that can be activated by not only exogenous pathogens and environmental irritants but endogenous danger signals and host-derived molecules as well [82]. Inflammasomes serve as platforms for caspase-1 mobilization and subsequent proteolytic maturation of the potent pro-inflammatory cytokine IL-1β. Inflammasome activation occurs when components localize to organelle clusters of mitochondria and endoplasmic reticulum (ER) [79, 83]. Tightly associated mitochondrial and ER membranes (MAMs) likely play an important role in inflammasome activation. Mitochondrial ROS contribute to this activation. Possibly, mitochondria release other factors involved in inflammasome activation (Fig. 11.2). An interesting trade-off was illuminated between mitophagy and inflammasome activity. Blocking mitophagy led to an accumulation of ROS-producing mitochondria and inflammasome activity [79, 83]. Tschopp [83] points out how these data resolve a long-standing paradox of inflammasome activity:

> …due to the multitude of danger signals sensed by the NLRP3 inflammasome, it has remained a mystery how a single molecule can achieve this almost impossible task. A plausible explanation is that instead of detecting each danger signal individually, the NLRP3 inflammasome monitors the activity of the mitochondrion, which acts as an integrator of danger signals, including those of metabolic origin.

The innate immune system of animals thus exhibits a number of constituents that connect to and interact with mitochondria. Connections with the mitochondria allow the immune system to carry out a central task of multicellularity, the coordinated repair mechanisms necessary to respond to various forms of damage.

Concluding Remarks If the history of life is a history of elaboration of levels of selection, numerous mechanisms of conflict mediation would have been deployed. Much complexity may have been built by co-opting these mechanisms either for new functions or for new conflicts as higher-level units arose. In both cases, vestiges

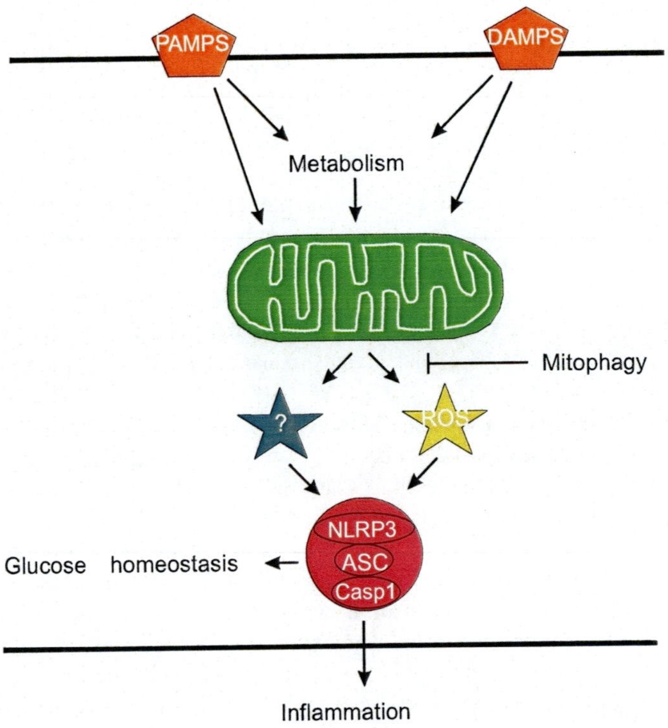

Fig. 11.2 The role of mitochondria in the activation of the inflammasome. Pathogen-associated molecular patterns (PAMPs) and damage-associated molecular patterns (DAMPs) directly or indirectly induce partial mitochondrial dysfunction. Mitochondrial ROS increase and possibly other factors are emitted (represented by "?"). Mitophagy removes these dysfunctional mitochondria and diminishes ROS. If left undiminished, via as-yet-uncharacterized mechanisms, ROS activate the inflammasome (NLRP3 + ASC + Casp1). An extracellular inflammatory response is triggered through cytokines IL-1β and IL-18. (Modified from Tschopp [83])

of past conflicts and their mediation should be apparent in molecular and cellular natural histories. Many of these vestiges should lead back to the mechanisms of chemiosmosis, which are suggested to be central to evolutionary conflict in previous chapters. The above examples, drawn from the literature on eukaryotic model systems of biomedical science, suggest the importance of metabolism, mitochondria, and reactive oxygen species to various signaling pathways (Fig. 11.3). These may

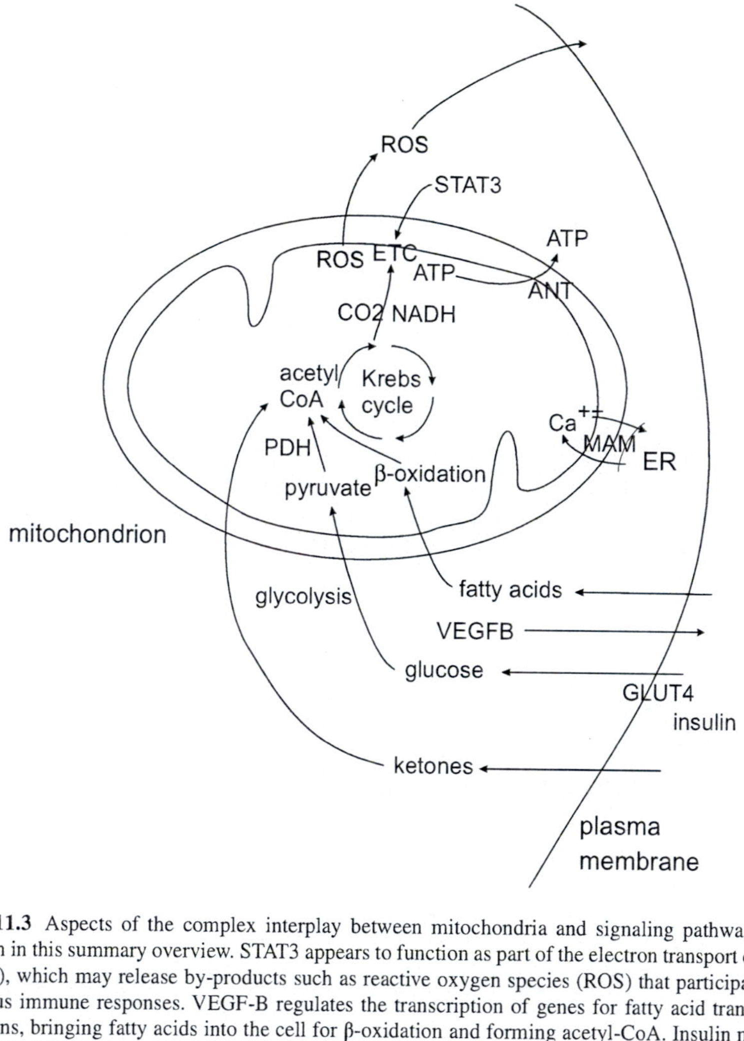

Fig. 11.3 Aspects of the complex interplay between mitochondria and signaling pathways as shown in this summary overview. STAT3 appears to function as part of the electron transport chain (ETC), which may release by-products such as reactive oxygen species (ROS) that participate in various immune responses. VEGF-B regulates the transcription of genes for fatty acid transport proteins, bringing fatty acids into the cell for β-oxidation and forming acetyl-CoA. Insulin mobilizes GLUT4 glucose transporters leading to pyruvate and acetyl-CoA via pyruvate dehydrogenase (PDH). Ketones also form acetyl-CoA, which is fed into the Krebs cycle, ultimately producing reducing equivalents for the electron transport chain. Calcium microdomains produced by the endoplasmic reticulum (ER) energize the oxidation of substrate by the electron transport chain via the mitochondria-associated ER membrane (MAM) and drive calcium signaling throughout the cell. These sorts of complex natural histories evolved in the context of the interplay between metabolism and evolutionary conflict and may represent vestiges of such conflict

be vestiges from the major transition that produced eukaryotes. Of course, there may be vestiges from major transitions that preceded the origin of eukaryotes. Indeed, metabolic signaling via the Krebs cycle might date back to the abiotic origins of life itself [84]. Of considerable interest would be the affinities of the genes involved in these signaling pathways. Generally, a functional dichotomy is found with archaeal (host) genes being involved in genetic information processing and bacterial (symbiont) genes being involved in metabolic processes [85]. Functional hybrids, such as those discussed above, might be included in either group.

Ultimately, the utility of any theory can be judged by the extent to which it provides useful predictions. In this regard, one prediction that can be offered is that more "mysterious" connections between signaling pathways and mitochondria remain to be discovered. Indeed, a profitable research program might entail searching specifically for such connections. In this way, the metabolic roots of cooperation and their more modern functions will increasingly be revealed.

References

1. Martin WF (2011) Early evolution without a tree of life. Biol Direct 6:36
2. Gould SJ, Vrba ES (1982) Exaptation—a missing term in the science of form. Paleobiology 8:4–15
3. Kluck RM, Bossy-Wetzel E, Green DR, Newmeyer DD (1997) The release of cytochrome c from mitochondria: a primary site for Bcl-2 regulation of apoptosis. Science 275:1132–1136
4. Yang J, Liu X, Bhalla K, Kim CN, Ibrado AM, Cai J, Peng TI, Jones DP, Wang X (1997) Prevention of apoptosis by Bcl-2: release of cytochrome c from mitochondria blocked. Science 275:1129–1132
5. Frade JM, Michaelidis TM (1997) Origin of eukaryotic programmed cell death—a consequence of aerobic metabolism. BioEssays 19:827–832
6. Kroemer G (1997) Mitochondrial implication in apoptosis: towards an endosymbiont hypothesis of apoptosis evolution. Cell Death Differ 4:443–456
7. Mignotte B, Vayssiere J-L (1998) Mitochondria and apoptosis. Eur J Biochem 252:1–15
8. Blackstone NW (2007) A food's-eye view of the transition from basal metazoans to bilaterians. Integr Comp Biol 47:724–733
9. Bredesen DE, Rao RV, Mehlen P (2006) Cell death in the nervous system. Nature 443:796–802
10. Blackstone NW, Green DR (1999) The evolution of a mechanism of cell suicide. BioEssays 21:84–88
11. Lane N (2005) Power, sex, suicide: mitochondria and the meaning of life. Oxford University Press, Oxford
12. Newman S (1994) Generic physical mechanisms of tissue morphogenesis: a common basis for development and evolution. J Evol Biol 7:467–488
13. Blackstone NW, Kirkwood TBL (2003) Mitochondria and programmed cell death: "slave revolt" or community homeostasis? In: Hammerstein P (ed) Genetic and cultural evolution of cooperation. MIT Press, Cambridge, MA, pp 309–325
14. Michod RE (2003) Cooperation and conflict mediation during the origin of multicellularity. In: Hammerstein P (ed) Genetic and cultural evolution of cooperation. MIT Press, Cambridge, MA, pp 291–307
15. Sablina AA, Budanov AV, Ilyinskaya GV, Agapova LS, Kravchenko JE, Chumakov PM (2005) The antioxidant function of the p53 tumor suppressor. Nat Med 11:1306–1313

16. Lu W-J, Chapo J, Roig I, Abrams JM (2010) Meiotic recombination provokes functional activation of the p53 regulatory network. Science 328:1278–1281
17. Rose MR (1991) Evolutionary biology of aging. Oxford University Press, Oxford
18. Nedelcu AM (2009) Comparative genomics of phylogenetically diverse unicellular eukaryotes provide new insights into the genetic basis for the evolution of the programmed cell death machinery. J Mol Evol 68:256–268
19. Vousden KH (2010) Alternative fuel—another role for p53 in the regulation of metabolism. Proc Natl Acad Sci U S A 107:7117–7118
20. Hu W, Zhang C, Sun Y, Levine A, Feng Z (2010) Glutaminase 2, a novel p53 target gene regulating energy metabolism and antioxidant function. Proc Natl Acad Sci U S A 107:7455–7460
21. Suzuki S, Tanaka T, Poyurovsky MV, Nagano H, Mayama T, Ohkubo S, Lokshin M, Hosokawa H, Nakayama T, Suzuki Y, Sugano S, Sato E, Nagao T, Yokote K, Tatsuno I, Prives C (2010) Phosphate-activated glutaminase (GLS2), a p53-inducible regulator of glutamine metabolism and reactive oxygen species. Proc Natl Acad Sci U S A 107:7461–7466
22. Tolstonog GV, Deppert W (2010) Metabolic sensing by p53: keeping the balance between life and death. Proc Natl Acad Sci U S A 107:13193–13194
23. Khutornenko AA, Roudko VV, Chernyak BV, Vartapetian AB, Chumakov PM, Evstafieva AG (2010) Pyrimidine biosynthesis links mitochondrial respiration to the p53 pathway. Proc Natl Acad Sci U S A 107:12828–12833
24. Bromberg J (2002) Stat proteins and oncogenesis. J Clin Invest 109:1139–1142
25. Levy DE, Darnell JE Jr (2002) Stats: transcriptional control and biological impact. Nat Rev Mol Cell Biol 3:651–662
26. Schindler CW (2002) JAK-STAT signaling in human disease. J Clin Invest 109:1133–1137
27. Levy DE, Lee C-K (2002) What does Stat3 do? J Clin Invest 109:1143–1148
28. Valsala H, Goshi N, Zhi-Xiang X, Fiona C, Hanausek M, Zbigniew W, Zoltaszek R, Gutterman JU (2009) Avicin D: a protein reactive plant isoprenoid dephosphorylates Stat 3 by both kinase and phosphatase activities. PLoS One 4:e5578
29. Gough DJ, Corlett A, Schlessinger K, Wegrzyn J, Larner AC, Levy DE (2009) Mitochondrial STAT3 supports Ras-dependent oncogenic transformation. Science 324:1713–1716
30. Myers MG Jr (2009) Moonlighting in mitochondria. Science 323:723–724
31. Wegrzyn J, Potla R, Chwae Y-J, Sepuri NBV, Zhang Q, Koeck T, Derecka M, Szczepanek K, Szelag M, Gornicka A, Moh A, Moghaddas S, Chen Q, Bobbili S, Cichy J, Dulak J, Baker DP, Wolfman A, Stuehr D, Hassan MO, Fu X-Y, Avadhani N, Drake JI, Fawcett P, Lesnefsky EJ, Larner AC (2009) Function of mitochondrial Stat3 in cellular respiration. Science 323:793–797
32. Blackstone NW (2009) Is evolutionary theory central to molecular cell biology? EvoS J J Evol Stud Consort 1:34–43
33. Embley TM, Martin W (2006) Eukaryotic evolution, changes and challenges. Nature 440:623–630
34. Michod RE, Nedelcu AM (2004) Cooperation and conflict during the unicellular-multicellular and prokaryotic-eukaryotic transitions. In: Moya A, Font E (eds) Evolution: from molecules to ecosystems. Oxford University Press, Oxford, pp 195–208
35. Martin WF, Garg S, Zimorski V (2015) Endosymbiotic theories for eukaryote origin. Philos Trans R Soc B 370:20140330
36. Martin WF (2017) Physiology, anaerobes, and the origin of mitosing cells 50 years on. J Theor Biol 434:2–10
37. Blackstone NW (1995) A units-of-evolution perspective on the endosymbiont theory of the orgin of the mitochondrion. Evolution 49:785–796
38. Georgellis D, Kwon O, Lin ECC (2001) Quinones as the redox signal for the arc two-component system of bacteria. Science 292:2314–2316
39. Allen JF (1993) Control of gene expression by redox potential and the requirement for chloroplast and mitochondrial genomes. J Theor Biol 165:609–631
40. Allen JF, Puthiyaveetil S, Ström J, Allen CA (2005) Energy transduction anchors genes in organelles. BioEssays 27:426–435

41. Pennisi E (2011) Green genomes. Science 332:1372–1375
42. Balic JJ, Albargy H, Luu K, Kirby FJ, Jayasekara WSN, Mansell F, Garama DJ, De Nardo D, Baschuk N, Louis C, Humphries F, Fitzgerald K, Latz E, Gough DJ, Mansell A (2020) STAT3 serine phosphorylation is required for TLR4 metabolic reprogramming and IL-1β expression. Nat Commun 11:3816. https://doi.org/10.1038/s41467-020-17669-5
43. Jacobson J, Duchen MR (2004) Interplay between mitochondria and cellular calcium signaling. Mol Cell Biochem 256(257):209–218
44. Berridge MJ, Bootman MD, Roderick HL (2003) Calcium signaling: dynamics, homeostasis and remodelling. Nat Rev Mol Cell Biol 4:517–529
45. Rizzuto R, Marchi S, Bonora M, Aguiari P, Bononi A, De Stefani D, Giorgi C, Leo S, Rimessi A, Siviero R, Zecchini E, Pinton P (2009) Ca2+ transfer from the ER to mitochondria: when, how, and why. Biochim Biophys Acta 1787:1342–1351
46. Carafoli E (2003) Historical review: mitochondria and calcium: ups and downs of an unusual relationship. Trends Biochem Sci 28:175–181
47. Chance B (1956) On possible mechanisms for the control of electron transport in the respiratory chain. In: Liebecqc C (ed) Proc 3rd intern Congr Biochem Brussels. Academic, pp 300–304
48. Rizzuto R, Brini M, Murgia M, Pozzan T (1993) Microdomains with high Ca2+ close to IP3-sensitive channels are sensed by neighboring mitochondria. Science 262:744–747
49. Blackstone NW (2015) The impact of mitochondrial endosymbiosis on the evolution of calcium signaling. Cell Calcium 57:133–139
50. Szabadkai G, Duchen MR (2008) Mitochondria: the hub of cellular Ca2+ signaling. Physiology 23:84–94
51. Hayashi T, Rizzuto R, Hajnoczky G, Su T-P (2009) MAM: more than just a housekeeper. Trends Cell Biol 19:81–88
52. De Stefani D, Raffaello A, Teardo E, Szabo I, Rizzuto R (2011) A forty-kilodalton protein of the inner membrane is the mitochondrial calcium uniporter. Science 476:336–340
53. Baughman JM, Perocchi F, Girgis HS, Plovanich M, Belcher-Timme CA, Sancak Y, Bao XR, Strittmatter L, Goldberger O, Bogorad RL, Koteliansky V (2011) Integrative genomics identifies MCU as an essential component of the mitochondrial calcium uniporter. Nature 476:341–345
54. Folkman J (2002) Role of angiogenesis in tumor growth and metastasis. Semin Oncol 29:15–18
55. Vogelstein B, Kinzler KW (2004) Cancer genes and the pathways they control. Nat Med 10:789–799
56. Tammela T, Enholm B, Alitalo K, Paavonen K (2005) The biology of vascular endothelial growth factors. Cardiovasc Res 65:550–563
57. Fischer C, Mazzone M, Jonckx B, Carmeliet P (2008) FLT1 and its ligands VEGFB and PlGF: drug targets for anti-angiogenic therapy? Nat Rev Cancer 8:942–956
58. Stefater JA III, Lewkowich I, Rao S, Mariggi G, Carpenter AC, Burr AC, Fan J, Ajima R, Molkentin JD, Williams BO, Wills-Karp M, Pollard JW, Yamaguchi T, Ferrara N, Gerhardt H, Lang RA (2011) Regulation of angiogenesis by a non-canonical Wnt-Flt1 pathway in myeloid cells. Science 474:511–515
59. Hagberg CE, Falkevall A, Wang X, Larsson E, Huusko J, Nilsson I, van Meeteren LA, Samen E, Lu L, Vanwildemeersch M, Klar J, Genove G, Pietras K, Stone-Elander S, Claesson-Welsh L, Yiä-Herttuala S, Lindahl P, Eriksson U (2010) Vascular endothelial growth factor B controls endothelial fatty acid uptake. Nature 464:917–921
60. Mehlem A, Palombo I, Wang X, Hagberg CE, Eriksson U, Falkevall A (2016) PGC-1α coordinates mitochondrial respiratory capacity and muscular fatty acid uptake via regulation of VEGF-B. Diabetes 65:861–873
61. Bonner JT (1998) The origins of multicellularity. Integr Biol 1:27–36
62. Brownlee M (2001) Biochemistry and molecular cell biology of diabetic complications. Nature 414:813–819
63. Simpson IA, Cushman SW (1986) Hormonal regulation of mammalian glucose transport. Annu Rev Biochem 55:1059–1089

64. Mukherjee C, Jungas RL (1975) Activation of pyruvate dehydrogenase in adipose tissue by insulin. Biochem J 148:229–235
65. Kashiwaya Y, Sato K, Tsuchiya N, Thomas S, Fell DA, Veech RL, Passonneau JV (1994) Control of glucose utilization in working perfused rat heart. J Biol Chem 269:25502–25514
66. Sato K, Kashiwaya Y, Keon CA, Tsuchiya N, King MT, Radda GK, Chance B, Clarke K, Veech RL (1995) Insulin, ketone bodies, and mitochondrial energy transduction. FASEB J 9:651–658
67. Kashiwaya Y, King MT, Veech RL (1997) Substrate signaling by insulin: a ketone bodies ratio mimics insulin action in the heart. Am J Cardiol 80:50A–64A
68. Hartman AL, Gasior M, Vining EPG, Rogawski MA (2007) The neuropharmacology of the ketogenic diet. Pediatr Neurol 36:281–292
69. Freeman JM, Kossoff EH, Hartman AL (2007) The ketogenic diet: one decade later. Pediatrics 119:535–543
70. Bough KJ, Rho JM (2007) Anticonvulsant mechanisms of the ketogenic diet. Epilepsia 48:44–58
71. Kashiwaya Y, Pawlosky R, Markis W, King MT, Bergman C, Srivastava S, Murray A, Clarke K, Veech RL (2010) A ketone ester diet increases brain malonyl-CoA and uncoupling proteins 4 and 5 while decreasing food intake in the normal Wistar rat. J Biol Chem 285:25950–25956
72. Wolkow CA, Kimura KD, Lee M-S, Ruvkun G (2000) Regulation of *C. elegans* life-span by insulin-like signaling in the nervous system. Science 290:147–150
73. Underhill DM, Ozinsky A (2002) Phagocytosis of microbes: complexity in action. Annu Rev Immunol 20:825–852
74. Delledonne M, Xia Y, Dixon RA, Lamb C (1998) Nitric oxide functions as a signal in plant disease resistance. Nature 394:585–587
75. Lam E, Kato N, Lawton M (2001) Programmed cell death, mitochondria and the plant hypersensitive response. Nature 411:848–853
76. Lambeth JD (2004) NOX enzymes and the biology of reactive oxygen species. Nat Rev Immunol 4:181–189
77. Sonoda J, Laganiere J, Mehl IR, Barish GD, Chong L-W, Li X, Scheffler IE, Mock DC, Bataille AR, Robert F, Lee C-H, Giguere V, Evans RM (2007) Nuclear receptor ERRα and coactivator PGC-1β are effectors of IFN-γ-induced host defense. Genes Dev 21:1909–1920
78. Harmata KL, Blackstone NW (2011) Reactive oxygen species and the regulation of hyperproliferation in a colonial hydroid. Physiol Biochem Zool 84:481–493
79. Zhou R, Yazdi AS, Menu P, Tschopp J (2010) A role for mitochondria in NLRP3 inflammasome activation. Nature 469:221–225
80. West AP, Brodsky IE, Rahner C, Woo DK, Erdjument-Bromage H, Tempst P, Walsh MC, Choi Y, Shadel GS, Ghosh S (2011) TLR signaling augments macrophage bactericidal activity through mitochondrial ROS. Nature 472:476–480
81. Vogel RO, Janssen RJRJ, van den Brand MAM, Dieteren CEJ, Verkaart S, Koopman WJH, Willems PHGM, Pluk W, van den Heuvel LPWJ, Smeitink JAM, Nijtmans LGJ (2007) Cytosolic signaling protein Ecsit also localizes to mitochondria where it interacts with chaperone NDUFAF1 and functions in complex I assembly. Genes Dev 21:615–624
82. Schroder K, Tschopp J (2010) The inflammasomes. Cell 140:821–832
83. Tschopp J (2011) Mitochondria: sovereign of inflammation? Eur J Immunol 41:1196–1202
84. Lane N (2009) Life ascending: the ten great inventions of evolution. Oxford University Press, Oxford
85. Brueckner J, Martin WF (2020) Bacterial genes outnumber archaeal genes in eukaryotic genomes. Genome Biol Evol 12:282–292

Chapter 12
Conclusions

> *Most biology departments include a group of faculty who study ecology, evolution and behavior as a single integrated subject, often labeled by an acronym such as EEB, and a group of faculty who study cell, biochemical, and molecular biology, often labeled by an acronym such as CBMB. Communication between these two groups is famously limited and occasionally even hostile.*
>
> *David Sloan Wilson et al. [1]*

As related in the previous chapters, two very different academic disciplines, typically segregated into sections within a department or even into separate departments, underwent revolutionary changes in the late twentieth century. Bioenergetics was roiled by the chemiosmotic hypothesis and subsequently the nearly two decades of "oxphos wars." During the same period, with some controversy, evolutionary biology for the first time seriously examined the circumstances under which cooperation could evolve while reaching the realization that, however unlikely, cooperation must have played a major role in the history of life.

Remarkably, the former scientific revolution may inform the latter. Chemiosmosis employs quantum electron transfer and other mechanisms that can rapidly accumulate products. Further, by separating hydrogen atoms into protons and electrons, chemiosmosis is a risky, fraught process. In the presence of molecular oxygen, stray electrons can rapidly form reactive oxygen species (ROS). ROS are widely used as biochemical signals, but at the same time can be perilous in large quantities. Since end-product inhibition can lead to ROS, a major challenge to a cell or organism—and one that parallels the conundrum faced by human societies' use of alternative energy—is consuming, storing, or simply getting rid of the products of chemiosmosis.

Thus, in many cases in modern biology, and at crucial junctures in the history of life, cells and organisms may have simply dispersed valuable chemiosmotic products into the environment. These products constitute usable substrate, and such substrate is the currency of evolutionary fitness. This "free lunch that you are forced to

N. W. Blackstone, *Energy and Evolutionary Conflict*,
https://doi.org/10.1007/978-3-031-06059-5_12

make" has led to groups in which other cells and organisms have taken advantage of this chemiosmotic largesse.

Since the time of Darwin, groups have been viewed as the key to cooperation. This perspective was reinforced with the rigorous development of the concepts of kin selection, reciprocal altruism, and group selection. All of these can lead to cooperation, and all of these involve groups either directly or indirectly. Further, once groups form, chemiosmosis can also enforce cooperation by punishing defectors that hoard products and in turn are incinerated by their own ROS.

No doubt, these ideas will be subject to a variety of criticisms. For instance, better mechanisms to regulate chemiosmosis could have evolved to eliminate waste. Certainly, there are mechanisms that alleviate the risk of end-product inhibition in chemiosmosis. Nevertheless, splitting hydrogen atoms into protons and electrons remains a fraught endeavor. Further, this may be looking at the problem backward. It may be precisely because chemiosmosis has produced so many successful collaborations in the history of life that a less well-regulated process has been favored and mechanisms of tight regulation have ultimately been selected against. Certainly, the two most successful symbioses in the history of life—those of mitochondria and plastids—were favored by less well-regulated chemiosmotic processes.

While all new ideas face criticism, in this case, the most likely response may be to simply ignore these ideas. Evolutionary biologists tend to dislike chemistry. This dislike often extends back to their days as undergraduate biology majors, when chemistry courses may have served as annoying "gatekeepers." On the other hand, perhaps these ideas can serve as an exemplar of the importance of chemistry even to fields that have become estranged from it. It may also be an opportune time for universities to consider the benefits of novel interdisciplinary programs such as evolutionary biochemistry. Certainly, one of the lessons of the current pandemic has been the power of natural selection and adaptation.

Ultimately, the value of these ideas will be apparent by their application. How many modern symbioses can be illuminated by reconsidering the costs and benefits of dispersing the products of chemiosmosis? For instance, on sunny days, plants provide mycorrhizal fungi with abundant photosynthate. Is this a cost or a benefit? Similarly, on sunny days, corals with symbiotic algae release large amounts of nutrient-rich material into the ocean, likely sustaining numerous other organisms. Is this a cost or a benefit? Chapter 8 provides other suggestions in this regard, but by-and-large these ideas are not current in the symbiosis literature. When defecting from the good of the multicellular group, mammalian cancer cells famously downregulate chemiosmosis when exhibiting the so-called Warburg effect. What are the consequences of these metabolic alterations? Of course, a large part of the foregoing discussion applies to more ancient events in the evolution of life. To what extent can the history of life be further illuminated by the simple principle of "follow the electrons?" We shall see.

Reference

1. Wilson DS, Geher G, Waldo J (2009) EvoS: Completing the evolutionary synthesis in higher education. EvoS J 1:3–10

Index

N. W. Blackstone, *Energy and Evolutionary Conflict*,
https://doi.org/10.1007/978-3-031-06059-5